STEM mindset™:
MATH *with* LEGO® *and* BRAINERS™

Grade 2
Book 2

Copyright © 2018 STEM mindset, LLC. All rights reserved.

STEM mindset, LLC 1603 Capitol Ave. Suite 310 A293 Cheyenne, WY 82001 USA
www.STEMmindset.com   info@stemmindset.com

STEM mindset™ and Brainers™ are the trademarks of STEM mindset, LLC.

LEGO® is a registered trademark of the LEGO Group which does not sponsor, endorse, or authorize this book.

The purchase of this material entitles the buyer to reproduce worksheets and activities for classroom use only – not for commercial resale. Reproduction of these materials for an entire school or district is strictly prohibited. No part of this book may be reproduced (except as noted before), stored in retrieval system, or transmitted in any form or by any means (mechanically, electronically, photocopying, recording, etc.) without the prior written consent of STEM mindset, LLC.

STEM mindset books are available at special discounts when purchased in bulk for premiums and sales promotions as well as for educational use.

First published in the USA 2018.   ISBN 978-1-948737-01-2

Table of Contents

CCSS Standards ....a

For Parents and Teachers ....d

Brainers ....k

Alphabet Table and Answers 163

Common Core State Standards

SKILLS in	Counting, Number & Operations	Algebraic thinking	Geometry	Measurement	Data Analysis, Statistics, & Probability
Addition	✓	✓	✓	✓	✓
Analyze Data	✓	✓	✓	✓	✓
Congruence	✓	✓	✓	✓	✓
Count	✓	✓	✓	✓	✓
Division	✓	✓	✓	✓	✓
Inequalities	✓	✓	✓	✓	✓
Language	✓	✓	✓	✓	✓
Manipulatives	✓	✓	✓	✓	✓
Match	✓	✓	✓	✓	✓
Pattern	✓	✓	✓	✓	✓
Shapes	✓	✓	✓	✓	✓
Subtraction	✓	✓	✓	✓	✓
Table	✓	✓	✓	✓	✓
Whole Numbers	✓	✓	✓	✓	✓

Math with Lego® and Brainers™ Grade 2 Book 2

SKILLS in	Counting, Number & Operations	Algebraic thinking	Geometry	Measurement	Data Analysis, Statistics, & Probability
Word Problems	✓	✓	✓	✓	✓
Coins	✓	✓		✓	✓
Estimate	✓		✓	✓	✓
Fraction	✓		✓	✓	✓
Money	✓	✓		✓	✓
Multiplication	✓	✓	✓		✓
Regrouping	✓	✓		✓	✓
Symmetry	✓	✓	✓	✓	
Bar/Column Graph	✓	✓			✓
Capacity	✓		✓	✓	
Denominator	✓		✓		✓
Length		✓	✓	✓	
More than 2 addends	✓		✓		✓
Number line	✓	✓	✓		
Numerator	✓		✓		✓
Probability	✓	✓			✓
Rounding	✓			✓	✓
Survey	✓	✓			✓
Time	✓			✓	✓
Line graph	✓				✓
Ruler			✓	✓	

SKILLS in	Counting, Number & Operations	Algebraic thinking	Geometry	Measurement	Data Analysis, Statistics, & Probability
Order of Operation	✓	✓			
Variable	✓	✓			
Calendar				✓	
Odd/Even	✓				
Place Value	✓				
Thermometer				✓	
Weight				✓	

MATH with LEGO and BRAINERS makes the foundations for students' success in MATH and Geometry:

- Common Core State Standards
- preparation for standardized tests such as SCAT®, CoGat®, and etc.
- full math and geometry curriculum that covers each grade level in great depth
- procedural fluency
- modeling and building with Lego bricks
- using and creating tables and diagrams for future programming skills
- numerous number sense strategies
- step-by-step strategies
- I-approach word problem strategies
- information coding (color-, border-, line coding)
- alive, engaging, and fun explanations with Brainers

- a good training for teachers in how to explain math concepts
- no lengthy teaching manuals for parents and teachers
- favorite children's font
- answers with step-by-step solutions

Our Approach Focuses on:

- all kids are gifted
- conceptual understanding
- reasoning, critical thinking skills
- flexibility
- creative-thinking skills
- STEM mindset growth
- visualization and play
- creating a foundation in how numbers work
- multi-step word problem strategies
- Olympiad math
- puzzles, brainteasers, game situations
- open, purposeful questions, -

to give the students the only treasure we can share – open a beautiful, amazing, adventurous, challenging world of mathematics and learning; the opportunity not only to do well on standardized tests but the opportunity to create the life-long learners; and to develop the traits which help our kids turn into happy and successful grown-ups.

For Parents and Teachers

When teaching a student, a common question arises: how can I help him or her develop critical, creative thinking? The student has no prior life experience, but from the first minutes of his/her life there are amazing abilities and needs to learn, explore, and investigate. How does a student learn new concepts?

There are two ways to success:

- memorization of the facts, ideas, methods, and new concepts, then, understanding these memorized concepts through visualization and real situations;

- visualization with real situations. These ideas generalize into concepts, strategies, and methods through reasoning, justifying, and proving. This method develops deeper conceptual understanding and thought-provoking learning.

Our aim is to awaken the student's interest, curiosity, and attention to mathematics as an important part of STEM mindset, to support independence in learning. Our rich experience confirms that it is possible if only the new concept has a common denominator with prior knowledge, creating harmony in the thinking. Math with Lego® and Brainers™ focuses on creating conceptual understanding through creative, critical, and problem-solving thinking, which eventually leads to procedural fluency.

Duke University in North Carolina described a well-known situation for a teacher. The students were given tests at the end of the year and they scored 95% and higher. After that they were given a one level higher test. Some students finished the next level test at 50%, while others reached only 5-20%. The difference is easily explained: The first group developed deeper conceptual understanding, which helped to apply procedural skills with greater fluency in new situations and reach better results.

Our Math is aimed at STEM mindset growth, and it includes many strategies and techniques to develop real STEM thinking, creativity, and curiosity in open STEM mindset students.

1. **The problems and curriculum in Math with Lego® and Brainers™ apply to Common Core State Standards**. The Common Core State Standards Table represents what kind of skills students will develop after learning. The series is not overwhelming and can be accomplished in a reasonable amount of time.

Our teaching experience has shown: the student can master one method perfectly, but if he is deprived of it, his former associations are destroyed, and he has nothing to rely on in the new concept. We offer a lot of counting, number sense, and word problem strategies, but each new strategy is introduced only when the prior strategy is worked out in different configurations using diagrams, tables, game situations, and visualization. As a result, each acquired strategy becomes the foundation for a new strategy.

2. **Visualization and Play develop deep understanding and open student thinking**. For example, if you ask a student to remember the definition of a rectangle as a shape consisting of 2 pairs of equal sides and four right angles, you can waste a lot of time on memorization. It would be much more effective if a student learned it through visualization and play (draw, color, take tooth picks, give a step-by-step strategy on how to make a rectangle with different kinds of sides). Visualization and Play contribute to the development of the inner imagination of students.

Visualization and play at this age helps to connect the manipulatives with one common feature - a number. Making deep connections between symbolic and visual representations of numbers helps a student get the concept of counting and reasoning.

3. **Active student engagement and involvement** (build, cut, write in, fill in, draw, break, make), and **investigation through open questions help students explore the math in an engaging and fun way**.

4. **Growth STEM mindset with an emphasis on MATH**. This strategy cultivates step-by-step processes from "low-level ceiling" problems to "high-level ceiling" problems in math, geometry, science, physics, geography, and programming. Each problem or activity is explained from the beginning. The so-called Olympiad math problems are relevant and interesting for the second-grade level of learning. STEM mindset is building on the usage of different methods and strategies to reinforce conceptual understanding.

STEM mindset growth is a long process, where we include numerous strategies. Thus, step-by-step number sense strategies encourage students to develop deeper levels of reasoning and critical thinking and make connections between concepts and facts.

5. **Open, purposeful questions** – where the answer depends on students' experience, knowledge, and critical thinking abilities – **encourage questioning and self-explanation**. These problems show that the only way to learn and find an answer is to investigate, think, reason, and use logic. They show that open-question does not demand only one correct answer. A student may make mistakes, struggle, and ask for help.

The problems also have help from "Brainers" (page k) which teach that it is fine to fear new math problems, make mistakes, and struggle with some problems which seem so simple and easy for others. We create these fears in our imaginary world and they prevent us from challenges, excitement, and adventure. When a student knows that anyone (a teacher or an adult) struggles with his own fears and makes his own mistakes, it is a way to successful learning for a student. Every baby challenges himself with turning, crawling, talking, and walking. Then, why do we think that they do not want to challenge themselves growing up with math? Let's show that each student, as much as an adult, has both adventurous, smart, kind, industrious, hard-working characteristics as well as grumpy, lazy, mad, angry, scared, and fearful.

6. **Information coding**:

- **color coding - highlighted (yellow) explanations and main ideas**. It helps students see the most important information, not just to read the rule or the mathematical explaining of theory. The engaging comical way helps students see what they need to pay more attention to;

- **border coding – all the numbers in problems are bordered**. It teaches the kids to pay attention to numeric information in the text;

- **line coding – all questions and commands are underlined**. It helps students pay attention to what they need to find, what they need to do.

Later these methods will come automatically, and teachers or parents will not need to remind each time to pay attention to main ideas, numbers, questions, or commands. (Yes, we are all tired of asking: What is the given information? What do you need to do? What are you asked about?) Reading the text, the students will automatically border numbers and underline commands or questions in the next grades.

7. **Brainers add humor to the math learning, making dull and boring math problems and explanations come alive, engaging, and fun**. Math with Lego® and Brainers™ does not require a teacher or an adult to read lengthy manuals on how to teach and explain new

concepts. The explanations made in a comical way by Brainers are applied to second-grade level reading. Students do not need guidance or assistance. The use of Brainers helps the student understand mathematics and geometry based on carefully selected critical and reasoning problems, but not on the explanations of a teacher or a parent. The Brainers' speeches have spaces for kids to write in the numbers or words. This engages a student in the learning and explaining process in a fun and interesting way. The problem often arises when students read theory explanations and they do not pay attention to what they are reading. After reading the usual question is: What was it about? Blank spaces make students read AND think.

8. **Why did we use the font** such as they see in their favorite children's books? If the child does not think independently, doesn't like to solve word problems, or he is bored with math, then it will be even worse and more boring if he reads the math assignments in strict adult kind of writing. He will be more likely to perceive the text he sees in his favorite books.

9. **Most of the word problems in our textbook are made with I-approach**. And this is not accidental. Any teacher or parent, explaining the problem at that age, says: "Imagine, you have … cubes (candy, toys), which you need to …." They lead a student to various mathematical activities through students' favorite topics: bricks, toys, cubes, candies, cookies, etc. This is the way that a student at this age perceives and understands the problem. It's no use to demand from a second-grader an answer to the problem like how many thousands of roses did *John* sell? The student learns through building, drawing, creating, making, playing, buying, eating - activities he is doing daily.

In addition, the word problems are followed with visualization:

- the use of tables, where students briefly record the main data. Later, they learn to select independently important data from the text;

- diagrams, which visually help the student to process and program information. Diagrams help the students develop the ability to break the task up into parts and thus, improve conceptual understanding.

10. **Use Lego (bricks)**. No secret, bricks are the most popular kind of toy for second-graders, and most schools are equipped with bricks for math classes. Students like to build and create. Implementing mathematics with their favorite plaything is the best solution for learning.

Even though there are children who are more interested in the issues of other academic subjects than math, we should not divide children into mathematicians and non-mathematicians. All children are gifted and capable, and all are capable of mathematics. Any subject which is fun, creative, engaging, and playful appeals to kids. If math is based on the same principles, it will be interesting to any student, too.

We talk a lot about the necessity to grow an open mindset, implement STEM principles into learning, go away from memorization, instruction, and assessment. As practical teachers, we know how difficult this aim is, how much work it demands, and we are not even sure about the results. Our Math with Lego® and Brainers™ is a tool to develop STEM mindset confidence and competence.

We are a dedicated team of teachers coming from a variety of teaching backgrounds across disciplines. Our teachers' credentials include Ph.Ds. and Masters in math, education, and social studies, and teaching experience of 20 plus years. We have brought together our vast and fulfilling teaching experience of kids from 4-year-olds to college students.

Contents: Number Sense strategies, Games, Word Problems Strategies, Adventure, Experiments, Olympiad math, Algorithms, Bricks, Puzzles, Brainteasers, Geometry, Patterning, Fun, Dream, Joy, Success.

We are Brainers. Yes, we live in the Brain. Some say we are thoughts, others insist we are ideas. Some demand we are emotions, others argue we are feelings. Some draw us as experience, others draw us as the future. We do not know who we are. We know that we exist in each Brain. We are what you think or feel when you open your eyes (or maybe when you close them,) when you do whatever you are doing, when you are asked to study, read, calculate, draw, write. We are different, and we want to open the mysterious door to the Wonderful, Amazing, Fantastic, Adventurous World of Math with you. Yes, you hear us right! Math is the BEST thing ever. Let's explore it together!

I'm the center of the brain (at least, I think so). I'm highly positive, adventurous, and open-minded. I'm ready to help or take any risk. I'm an encouraging and unbelievably positive friend.

I'm the Brain's disrespectful, grumpy, bad-tempered, critical Brainer. I get "crazy-mad", impatient with "any problem", or displeased with "any explanation". I'm often quarrelsome and disagreeable with anything new.

I'm "the-smartest-ever-lived", decisive, and loyal to any problem and any Brainer. I'm sharp, and competent in everything. My common sense can explain any problem or calm down any argument between Brainers. I'm "hard-working and extremely ambitious". So, I'm "too brainy".

I'm very occupied with fear of everything, especially new. I'm the terror of any problem, especially word problems. I get panicky and terrified by many pages of the workbook. I'm a "scaredy-cat" according to some critical members of the Brain but encouraged by others.

I'm enthusiastic, excited, and sure about everything in this wonderful world. I'm extremely good-natured and trusting. I'm often engaged in any activity as I'm enterprising, industrious and cheerful. I'm "too energetic and happy" according to some Brainers.

I love to be "left in peace and alone". I would be a dreamer if not for other Brainers and math. Sometimes I'm ready to give up right away or resist anything new..."WHY should I???". I'm persistent in doing nothing.

Good luck! Let's start our math adventure!

Yes, my friend, you are right. I'd rather play football, but...college, bills, you know... Math is everywhere...

Math is fun, playful, and beautiful!

Help! Help! Help! I need help with all math...

Don't worry! We will help!

Let's start, then. The sooner the better, right?

1. <u>Answer</u> the questions and <u>fill in</u> the missing letters.

Below is a ⟦7-inch⟧ long rectangle divided into ⟦7 equal⟧ parts.

Hints: 1) <u>draw</u> a ⟦7-inch⟧ long rectangle; 2) <u>measure</u> ⟦7 equal⟧ parts of ⟦1-inch⟧ long; 3) <u>draw</u> the 7 vertical black-dotted lines. You will get ⟦7⟧ equal parts.

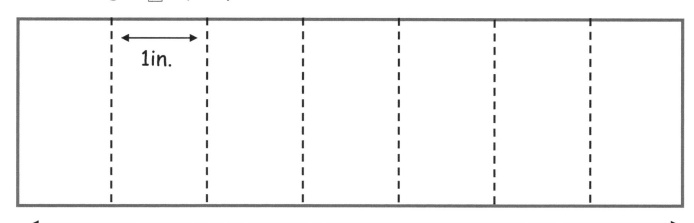

<u>How many</u> is ⟦1⟧ part out of ⟦7⟧ equal parts? $1 \div 7 = \frac{1}{7}$ (one-seventh).

<u>How many</u> are ⟦2⟧ parts out of ⟦7⟧? $\ldots \div \ldots = \frac{\ldots}{\ldots}$ (_____).

<u>How many</u> are ⟦3⟧ parts out of ⟦7⟧? $\ldots \div \ldots = \frac{\ldots}{\ldots}$ (_____).

<u>How many</u> are ⟦4⟧ parts out of ⟦7⟧? $\ldots \div \ldots = \frac{\ldots}{\ldots}$ (_____).

<u>How many</u> are ⟦5⟧ parts out of ⟦7⟧? $\ldots \div \ldots = \frac{\ldots}{\ldots}$ (_____).

<u>How many</u> are ⟦6⟧ parts out of ⟦7⟧? $\ldots \div \ldots = \frac{\ldots}{\ldots}$ (_____).

<u>How many</u> are ⟦7⟧ parts out of ⟦7⟧? $\ldots \div \ldots = \frac{\ldots}{\ldots} = 1$.

The whole always equals 1 or $\frac{2}{2}$, or $\frac{3}{3}$, or $\frac{4}{4}$, or $\frac{6}{6}$, or $\frac{7}{7}$.

1. Number Sense Strategy. Fill in the missing numbers to complete each number sentence. Use the bricks. Circle the bricks to show the difference. You can subtract the leftover bricks in any order and quantity.

15 – 7 = 8

Aha... I circle 8 bricks (the difference), and the 7 bricks that are leftover, I arrange as I want.

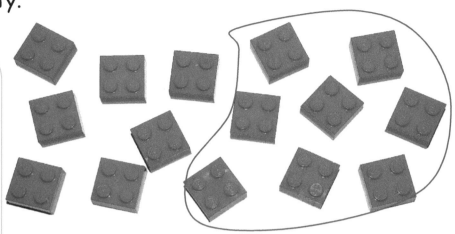

15 - ... - ... = 8   15 - ... - ... = 8
15 - ... - ... = 8   15 - ... - ... = 8
15 - ... - ... - ... = 8   15 - ... - ... - ... = 8
15 - ... - ... - ... = 8   15 - ... - ... - ... - ... = 8
15 - ... - ... - ... - ... = 8   15 - ... - ... - ... - ... - ... = 8

2. I have played 12 games in badminton. My brother has won 5 games. How many games have I won?

Played	Brother won	I won
...	...	? ...

Won: ...   Won: ? ...

Played: ...

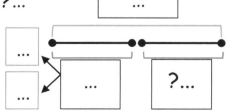

Answer: ___

1. <u>Cut out</u> 9 equal squares of 3 colors: 3 – pink squares, 3 – green squares, 3 – yellow squares.

   So, you have 9 squares of 3 colors.

   Now <u>make</u> 1 big square with 9 small squares.

   <u>Arrange</u> the squares so that each row, each column, and 1 numeral diagonal of the 9-squared shape has 3 different colors and 1 numeral diagonal has 3 equal colors.

2. <u>Follow</u> the directions.

   <u>Draw</u> the minute and hour hands red for:

   7:00 am. <u>Draw</u> the minute and hour hands black and <u>fill in</u> the missing numbers and letters for: 5 and a half hours later (…:…… _____).

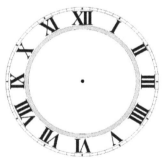

3. <u>Write</u> the numbers between:

   122   … …   125   539   … …   542
   483   … …   486   890   … …   893
   248   … …   251   965   … …   968

1. Number Sense Strategy. Write subtraction number sentences for each picture. Circle the bricks by 10's. Cross out the number of bricks you subtract. The first one is done for you.

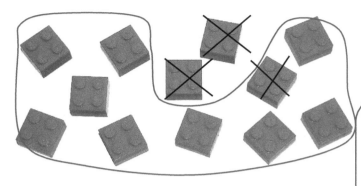

12 - 3 = ...
   ↙ ↘
  2   1   12 - 2 - 1 = ...

I like to count by 10's, so, how to make 10 out of 12? Subtract 2: 12 - 2. And then, subtract the remaining 1.

13 - 9 = ...
   ↙ ↘
  ... ...   13 - ... - ... = ...

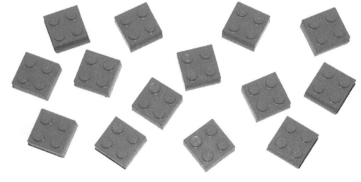

14 - 9 = ...
   ↙ ↘
  ... ...   14 - ... - ... = ...

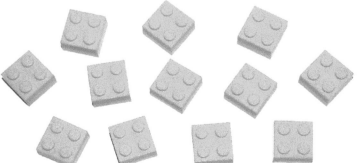

12 - 4 = ...
   ↙ ↘
  ... ...   12 - ... - ... = ...

1. Number Sense Strategy. Fill in the missing numbers to complete each number sentence. Use the bricks. The bricks are circled to show the difference. You can subtract the leftover bricks in any order and quantity.

16 − 7 = 9

16 − ... − ... = 9

16 − ... − ... = 9

16 − ... − ... = 9

16 − ... − ... = 9

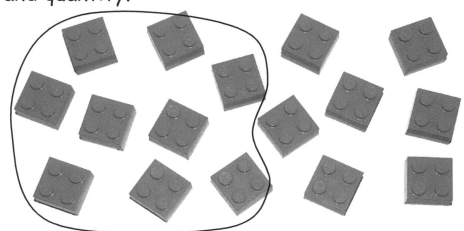

16 − ... − ... − ... = 9   16 − ... − ... − ... = 9

16 − ... − ... − ... = 9   16 − ... − ... − ... − ... = 9

16 − ... − ... − ... − ... = 9   16 − ... − ... − ... − ... − ... = 9

2. If I do $\frac{2}{8}$ of my homework per day, how much homework will be done in 4 days?

Per day	Days	Homework
$\frac{...}{...}$	...	?$\frac{...}{...}$

$1 = \frac{...}{...}$

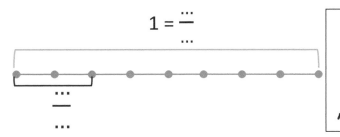

$\frac{...}{...}$

Answer:_____

3. Answer the questions.

4972: the sum of the ones and hundreds is _____. The difference between the tens and thousands is _____.

1. <u>Answer</u> the questions. <u>Fill in</u> the missing numbers and words.

I had ☐1 whole apple and a ☐half ($\frac{1}{2}$) of an apple. I shared $\frac{1}{2}$ of an apple with a friend. <u>How many</u> is left?

You can <u>choose</u> one of two strategies:

$1\frac{1}{2} - \frac{1}{2} = \frac{...}{...} - \frac{...}{...} = \frac{...\,-\,...}{...} = \frac{...}{...}$ or _____.

or

$1\frac{1}{2} - \frac{1}{2} = ...$ 1) $\frac{...}{...} - \frac{...}{...} = \frac{...\,-\,...}{...} = ...$ 2) ... - ... = ...

2. <u>What</u> is the ☐smallest possible 3-digit number if each digit is different?   ... ...

<u>What</u> is the ☐biggest possible 3-digit number if each digit is different?   ... ...

3. <u>Find out</u> the numbers hiding behind the bricks. <u>Complete</u> the number sentences.

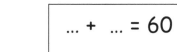
... + ... = 60

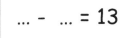
... − ... = 13

... + ... = 38

We need to find the value of 11 – 3. Who has any idea?

11 – 3 = …

Yesterday I was reading a book and they gave a very interesting explanation. I liked it as I like to calculate in tens:

11 is 10 + 1. So, 10 – 3 = 7. Then, I add: 7 + 1 = 8.

I want to subtract 15 – 8. Help me!

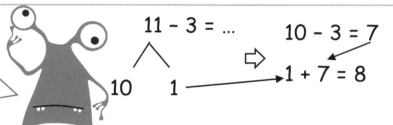

15 is 10 + 5. So, subtract: 10 – 8 = 2.

Last, add 5 + 2.

15 – 8 = …   10 – 8 = 2
  /\   5 + 2 = …
10   5

1. <u>Fill in</u> the missing numbers.

<u>Find</u> the missing of 2 odd consecutive numbers.   …,   13.

<u>Find</u> the missing of 3 odd consecutive numbers.   37,   …,   ….

<u>Find</u> the missing of 4 odd consecutive numbers.   …,   …,   25,   ….

<u>Find</u> the missing of 5 odd consecutive numbers.   …, …, …, …, 11.

1. Number Sense Strategy. Write subtraction number sentences for each picture. Circle the bricks by 10's. Cross out the number of bricks you subtract.

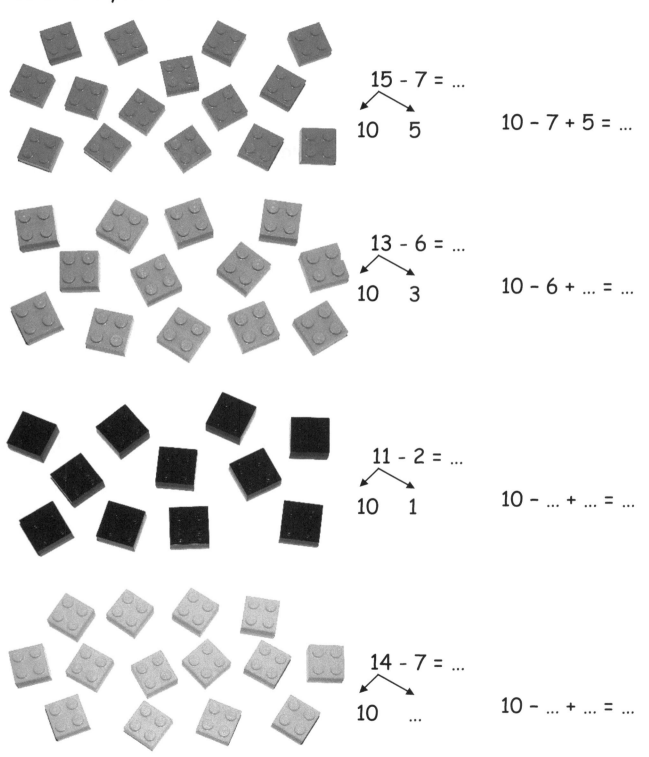

1. Number Sense Strategy. Fill in the missing numbers to complete each number sentence. Use the bricks. Circle the bricks to show the difference. You can subtract the leftover bricks in any order and quantity.

11 − 8 = 3

11 − ... − ... = 3

11 − ... − ... = 3

11 − ... − ... = 3

11 − ... − ... = 3

11 − ... − ... − ... = 3   11 − ... − ... − ... = 3

11 − ... − ... − ... = 3   11 − ... − ... − ... − ... = 3

11 − ... − ... − ... − ... = 3   11 − ... − ... − ... − ... − ... = 3

2. Draw the minute and hour hands red for midnight.

How much time is left until 5 past 6 am?

    ... hours ... minutes

Draw the minute and hour hands black.

3. Today is Tuesday, the 12th. If I celebrate my birthday in 13 days, what day will it be (day of the week and date)?

_____.

1. We are waiting for ⬚17⬚ guests. I put ⬚9⬚ forks and ⬚5⬚ spoons on the table. <u>How many more forks and spoons</u> do I need?

Guests	Forks	Spoons	More forks	More spoons
...	...	...	? ...	? ...

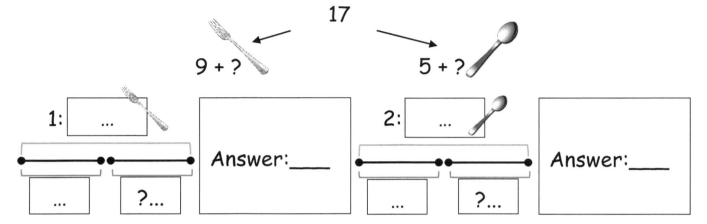

2. NSS. <u>Find</u> the value. <u>Do you see</u> a pattern in these problems?

| 13 − 4 + 5 = ... |
| 13 − 5 + 4 = ... |

| 11 − 3 + 4 = ... |
| 11 − 4 + 3 = ... |

| 12 − 4 + 5 = ... |
| 12 − 5 + 4 = ... |

| 12 − 3 + 4 = ... |
| 12 − 4 + 3 = ... |

| 14 − 4 + 5 = ... |
| 14 − 5 + 4 = ... |

| 11 − 4 + 5 = ... |
| 11 − 5 + 4 = ... |

I noticed that _____

_____

_____.

1. <u>Answer</u> the questions and <u>fill in</u> the missing letters.

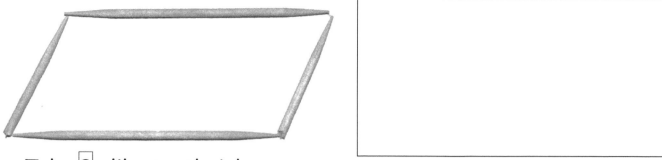

<u>Take</u> 3 alike toothpicks.

<u>Break</u> 1 toothpick in the middle into 2 equal parts.

<u>Connect</u> the ends of the 2 broken parts with the 2 whole toothpicks as it's shown in the picture. The 2 opposite sides are equal and parallel.

<u>Sketch</u> your shape in the box.

<u>How many sides</u> does it have?   … sides.

<u>How many angles</u> does it have?   … angles.

A shape with 2 pairs of equal (or congruent) sides, which are parallel, and 4 angles is called

… … … … … … … … … … … … …

16 1 18 1 12 12 5 12 15 7 18 1 13

(to solve this puzzle, <u>look</u> at the back of the book and <u>find</u> out what place each letter takes in the alphabet: A is 1st, Z is 26th).

<u>How many toothpicks</u> do you need to make a parallelogram if one pair of sides has 2 toothpicks each side and another pair of sides has 3 toothpicks each side? _____.

1. Number Sense Strategy. Write subtraction number sentences for each picture. Circle the bricks by 10's. Cross out the number of bricks you subtract.

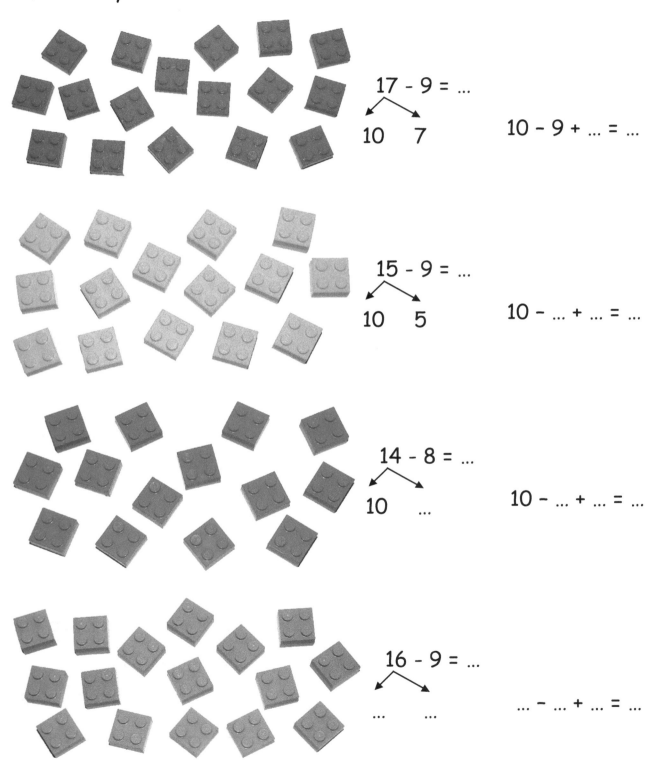

17 − 9 = ...
10   7

10 − 9 + ... = ...

15 − 9 = ...
10   5

10 − ... + ... = ...

14 − 8 = ...
10   ...

10 − ... + ... = ...

16 − 9 = ...
...   ...

... − ... + ... = ...

1. Number Sense Strategy. Fill in the missing numbers to complete each number sentence. Use the bricks. Circle the bricks to show the difference. You can subtract the leftover bricks in any order and quantity.

12 − 8 = 4

12 - ... - ... = 4

12 - ... - ... = 4

12 - ... - ... = 4

12 - ... - ... = 4

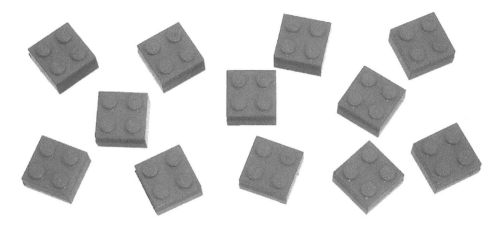

12 - ... - ... - ... = 4   12 - ... - ... - ... = 4

12 - ... - ... - ... = 4   12 - ... - ... - ... - ... = 4

12 - ... - ... - ... - ... = 4   12 - ... - ... - ... - ... - ... = 4

2. I wrote 15, 3, 8, 18, 11, 4 on the blackboard. How many more is the largest number than the smallest number?

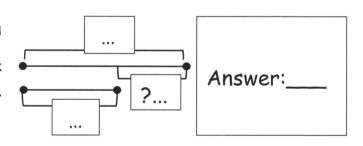

Answer:___

3. Fill in the missing numbers. Find the value. Compare (> or <).

$1 - \frac{5}{6} = \frac{...}{...} - \frac{...}{...} = \frac{...-...}{...} = \frac{...}{...}$ ... $1 - \frac{2}{3} = ... \frac{...}{...} - \frac{...}{...} = \frac{...-...}{...} = \frac{...}{...}$

1. Fill in the missing numbers.

Hundreds	Tens	Ones	Number
8	9	4	...
3	7	1	...
7	4	9	...
4	6	2	...
...	...	...	632
...	...	...	879
...	...	...	126
...	...	...	504
...	...	...	295

2. Ask an adult for help to measure 1 pound of beans, peas, apples, and oranges. Find out how many of each fruit or vegetable you have in 1 pound. You can change the fruits or the vegetables.

1 pound of beans equals _____ beans.

1 pound of peas equals _____ peas.

1 pound of apples equals _____ apples.

1 pound of oranges equals _____ oranges.

Measure your weight and write it in: ... pounds;

...kilograms.

Measure 1 cup of water with 2 tablespoon of salt:... ounces;

... grams.

Measure 1 cup of pure water: ... ounces;

... grams.

1. I drew 5 line segments. Measure the line segments in inches.

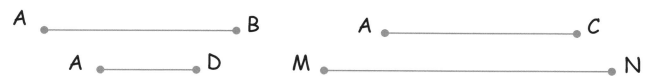

Add the lengths and write how many inches they make together according to the letters below. Draw the line segments you've got on the grid.

AB + AC = KL = ... + ... = ...   AD + AB = NL = ... + ... = ...

MN + AD = ML = ... + ... = ...   AC + AD = QR = ... + ... = ...

AB + MN = PR = ... + ... = ...   AC + MN = OP = ... + ... = ...

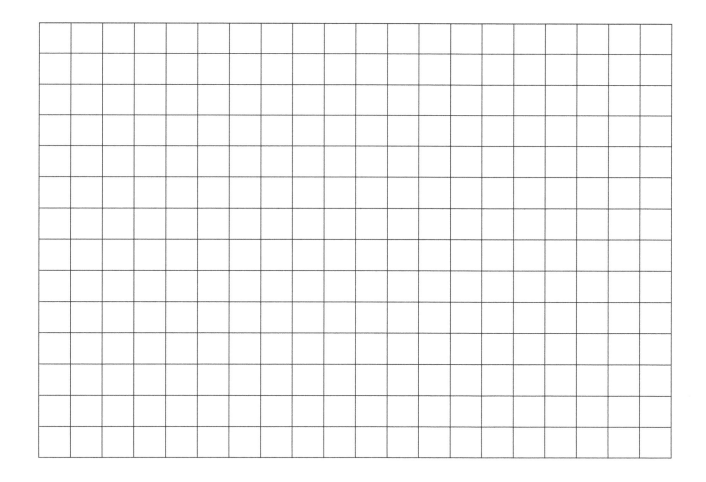

1. Number Sense Strategy. Write subtraction number sentences for each picture. Circle the bricks by 10's. Cross out the number of bricks you subtract.

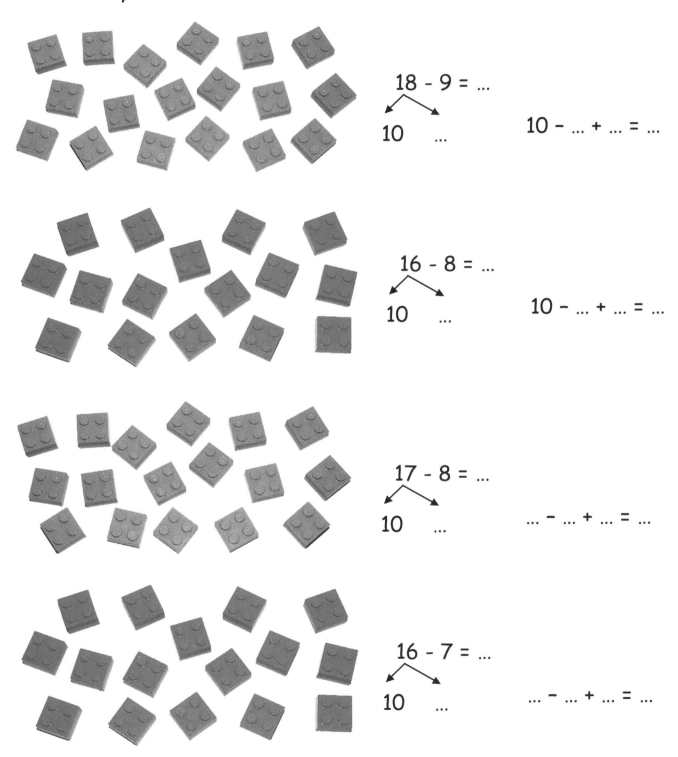

1. Number Sense Strategy. Fill in the missing numbers to complete each number sentence. Use the bricks. Circle the bricks to show the difference. You can subtract the leftover bricks in any order and quantity.

13 – 8 = 5

13 - ... - ... = 5

13 - ... - ... = 5

13 - ... - ... = 5

13 - ... - ... = 5

13 - ... - ... - ... = 5

13 - ... - ... - ... = 5

13 - ... - ... - ... = 5

13 - ... - ... - ... - ... = 5

13 - ... - ... - ... - ... = 5

2. I wrote 21, 5, 27, 11, 9 on the blackboard. How many more is the largest number than the smallest number?

Answer:____

3. My 3 friends and I have played 6 chess games. Each has played an equal number of games. How many games have I played? Use the diagram to solve the problem.

_____.

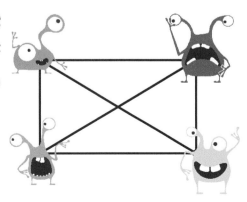

1. Number Sense Strategy. Write subtraction number sentences for each picture. Use the strategy of rounding up the subtrahend.

12 - 7 = …
10   3

12 – 10 + 3 = …

Rounding up? In subtraction? Mom, help!!!

We need to round up the subtrahend 7. It's 10.
Then, what is 7?
7 = 10 - 3.
So, 12 – 10 + 3 = ….

13 - 4 = …
10   6

13 – 10 + … = …

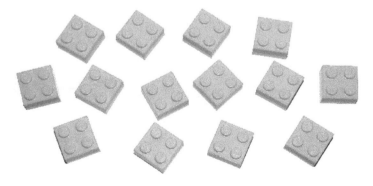

Subtrahend
14 - 6 = …
10   …

14 – … + … = …

11 - 6 = …
…   …

11 – … + … = …

1. <u>Answer</u> the questions and <u>fill in</u> the missing letters.

Why does one end go up?..

I've used toothpicks!!!

<u>Take</u> 4 alike toothpicks.

<u>Break</u> 1 toothpick in the middle into 2 equal parts and <u>break another</u> toothpick into 2 unequal parts, <u>throw away</u> the smaller part.

<u>Connect</u> the ends of the 3 broken parts with 1 whole toothpick as it's shown in the picture. 2 sides are equal and not parallel. 2 other sides are unequal, but parallel.

<u>Sketch</u> your shape in the box.

<u>How many sides</u> does it have? … sides.

<u>How many angles</u> does it have? … angles.

A shape with at least 1 pair of parallel sides (which are called the base) and 4 angles is called

… … … … … … … … …
20 18 1 16 5 26 15 9 4

(to solve this puzzle, <u>look</u> at the back of the book and <u>find</u> out what place each letter takes in the alphabet: A is 1ˢᵗ, Z is 26ᵗʰ).

1. Number Sense Strategy. Fill in the missing numbers to complete each number sentence. Use the bricks. Circle the bricks to show the difference. You can subtract the leftover bricks in any order and quantity.

14 – 8 = 6

14 - ... - ... = 6

14 - ... - ... = 6

14 - ... - ... = 6

14 - ... - ... = 6

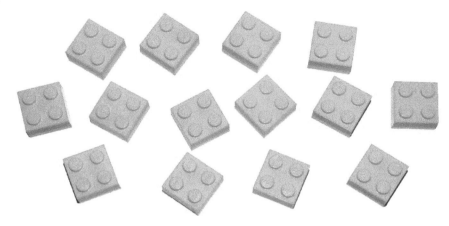

14 - ... - ... - ... = 6   14 - ... - ... - ... = 6

14 - ... - ... - ... = 6   14 - ... - ... - ... - ... = 6

14 - ... - ... - ... - ... = 6   14 - ... - ... - ... - ... - ... = 6

2. Draw the minute and hour hands red for: 8:20 am. My school is over at 1:30pm. How much time do I spend at school?
_____.

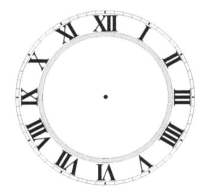

3. Do you know that any month starting on Sunday, has Friday, the 13th? Check it on the calendar and write the months which have Friday the 13th this year: _____.

1. Number Sense Strategy. Write subtraction number sentences for each picture. Use the strategy of rounding up the subtrahend.

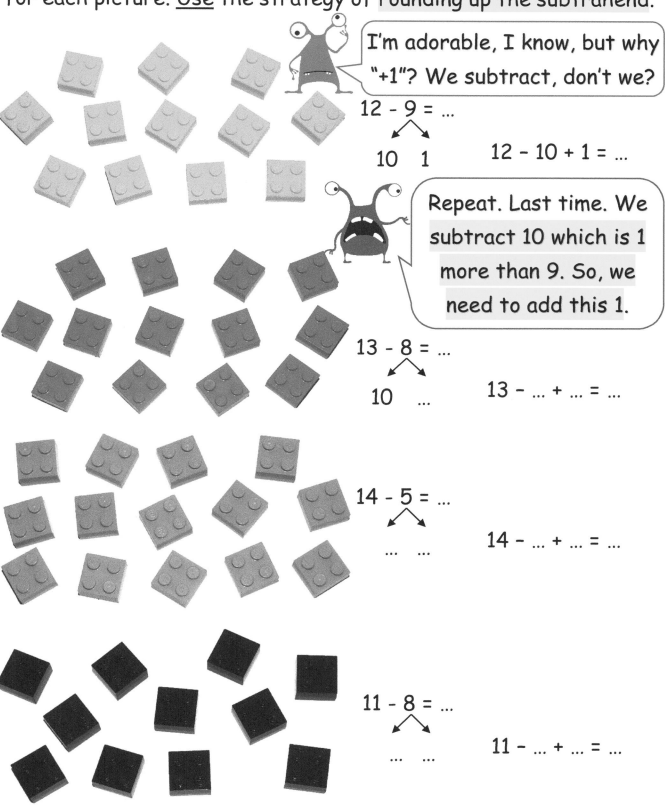

"I'm adorable, I know, but why "+1"? We subtract, don't we?"

12 - 9 = ...
  10   1
12 − 10 + 1 = ...

"Repeat. Last time. We subtract 10 which is 1 more than 9. So, we need to add this 1."

13 - 8 = ...
  10   ...
13 − ... + ... = ...

14 - 5 = ...
  ...   ...
14 − ... + ... = ...

11 - 8 = ...
  ...   ...
11 − ... + ... = ...

1. 8 kids came to my birthday party at 2 pm, 6 more kids came at 2:30 pm. 7 kids have to leave at 4:30 pm. How many kids stayed? Solve the problem with two number sentences and then, with one number sentence.

Kids at 2:00	Kids at 2:30	Kids in all	Left at 4:30	Stayed
...	...	? ...	...	? ...

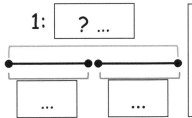

1: ? ...   Answer:___

2: ...   ? ...   Answer:___

Two number sentences: ... + ... = ... - ... = ...

One number sentence: (... + ...) - ... = ...

The first step is in parenthesis ().

2. What bills and coins may make up $8.99? Fill in the missing bills and coins.

1 5-dollar bill, ... 1-dollar bill, 2 quarters, ... dimes, ... nickels, 39 pennies.

... 5-dollar bill, 1 1-dollar bill, ... quarters, ... dimes, 30 nickels, 19 pennies.

3. How many angles do you see here? _____.

What kinds of angles are they? _____.

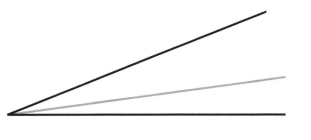

1. Number Sense Strategy. Fill in the missing numbers to complete each number sentence. Use the bricks. Circle the bricks to show the difference. You can subtract the leftover bricks in any order and quantity.

15 – 8 = 7

15 - ... - ... = 7

15 - ... - ... = 7

15 - ... - ... = 7

15 - ... - ... = 7

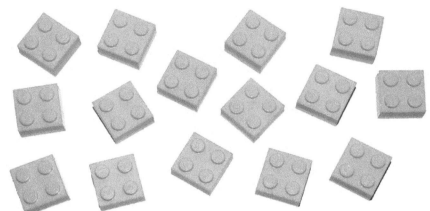

15 - ... - ... - ... = 7   15 - ... - ... - ... = 7

15 - ... - ... - ... = 7   15 - ... - ... - ... - ... = 7

15 - ... - ... - ... - ... = 7   15 - ... - ... - ... - ... - ... = 7

2. Write the numbers.

_____ four hundred seventy-four

_____ seven hundred forty-eight

_____ six hundred twenty-six

_____ nine hundred sixty-nine

3. Take 3 equal toothpicks and connect the ends to make a triangle. Draw what you got in the box.

1. <u>Fill in</u> the missing numbers and <u>find</u> the value. The first one is done for you.

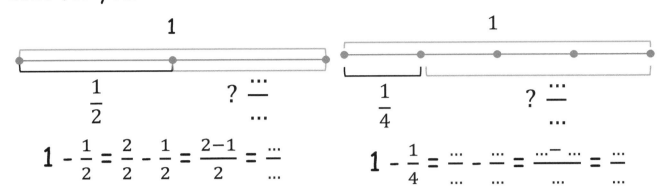

$1 - \dfrac{1}{2} = \dfrac{2}{2} - \dfrac{1}{2} = \dfrac{2-1}{2} = \dfrac{...}{...}$

$1 - \dfrac{1}{4} = \dfrac{...}{...} - \dfrac{...}{...} = \dfrac{...-...}{...} = \dfrac{...}{...}$

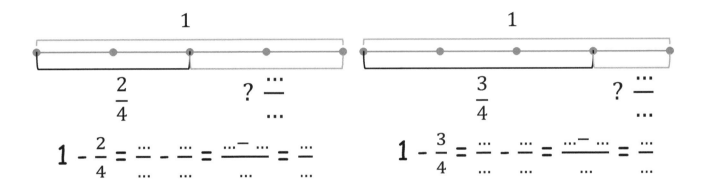

$1 - \dfrac{2}{4} = \dfrac{...}{...} - \dfrac{...}{...} = \dfrac{...-...}{...} = \dfrac{...}{...}$

$1 - \dfrac{3}{4} = \dfrac{...}{...} - \dfrac{...}{...} = \dfrac{...-...}{...} = \dfrac{...}{...}$

2. <u>Write</u> the names of the months.

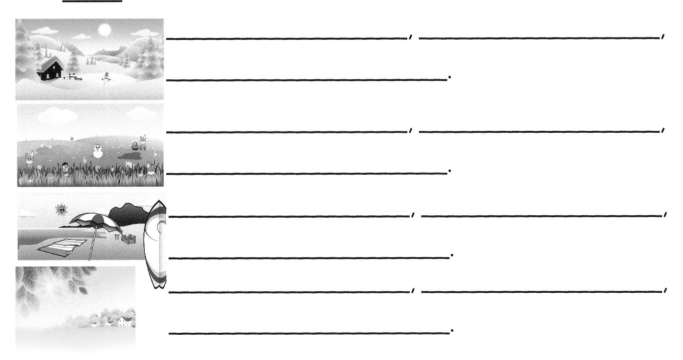

_____, _____,
_____.

_____, _____,
_____.

_____, _____,
_____.

_____, _____,
_____.

1. **Number Sense Strategy.** <u>Write</u> subtraction number sentences for each picture. <u>Use</u> the strategy of rounding up the subtrahend.

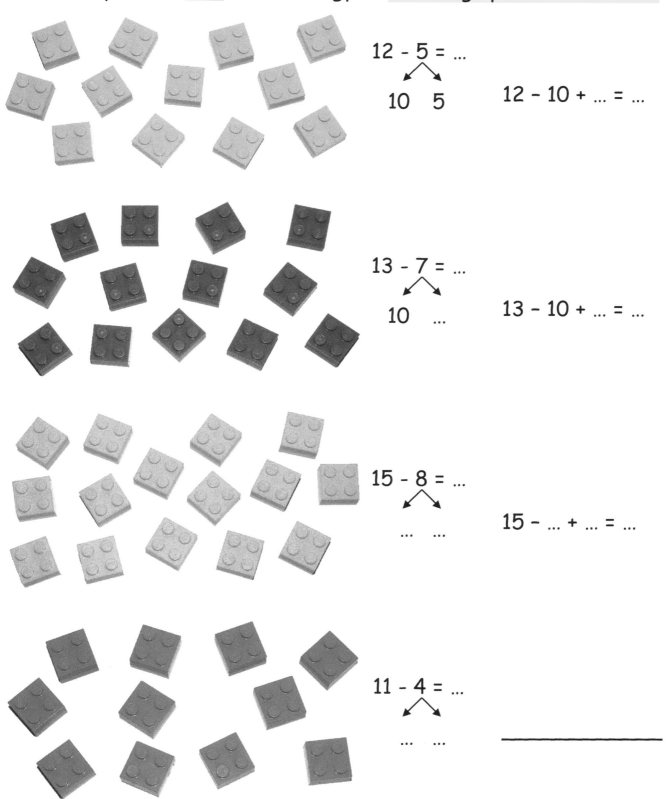

1. I invited ⌐15¬ kids to my birthday. My sister put ⌐9¬ plates (Pl.), ⌐7¬ spoons (Sp.) and ⌐6¬ cups(C.) on the table. How many more plates, spoons and cups does she need? Draw the diagrams.

Kids	Plates	More Plates	Spoons	More Spoons	Cups	More Cups
...	...	?...	...	?...	...	?...

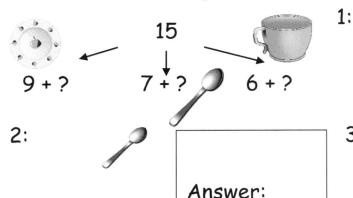

1:

Answer:___

2:

Answer:___

3:

Answer:___

2. Answer the questions.

⌐9564¬: the sum of the ones and hundreds is _____, the sum of the ones and thousands is _____. The difference between the thousands and tens is _____.

3. Take ⌐3 unequal¬ toothpicks of lengths: a ⌐whole, a half, and one-fourth¬ and connect the ends to make a triangle. Draw what you got in the box. Why can or can't you?

_____.

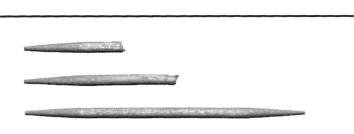

1. Number Sense Strategy. Insert the missing numbers to complete each number sentence. Use the bricks. Circle the bricks to show the difference. You can subtract the leftover bricks in any order and quantity.

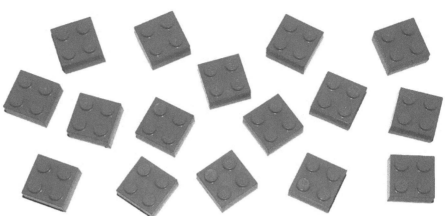

16 – 8 = 8

16 - ... - ... = 8

16 - ... - ... = 8

16 - ... - ... = 8   16 - ... - ... = 8

16 - ... - ... - ... = 8   16 - ... - ... - ... = 8

16 - ... - ... - ... = 8   16 - ... - ... - ... - ... = 8

16 - ... - ... - ... - ... = 8   16 - ... - ... - ... - ... - ... = 8

2. My bedtime is 8:45pm. Draw the minute and hour hands red.

If I wake up at 6:35am, how long am I sleeping?

_____.

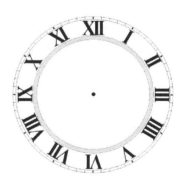

3. How many times do the hour and minute hands pass each other in 2 hours from 1:15pm to 3:15pm.? Write the time(s).

_____.

1. Circle the shapes which are called trapezoids. Write the names of the shapes.

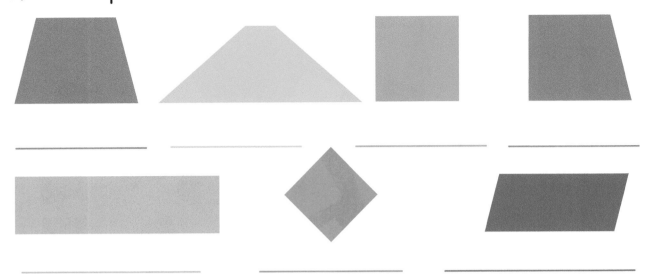

2. 8 kids came to my birthday party at 1 pm. 8 more kids came at 1:30 pm. 9 kids have to leave at 3:30pm. How many kids stayed? Solve the problem with two number sentences and then, with one number sentence. Draw the diagrams.

Kids at 1:00	Kids at 1:30	Kids in all	Left at 3:30	Stayed
...	...	? ...	...	? ...

1:

Answer:___

2:

Answer:___

Two number sentences: _____

One number sentence: _____

1. Number Sense Strategy. Write subtraction number sentences for each picture. Use the strategy of rounding up the subtrahend.

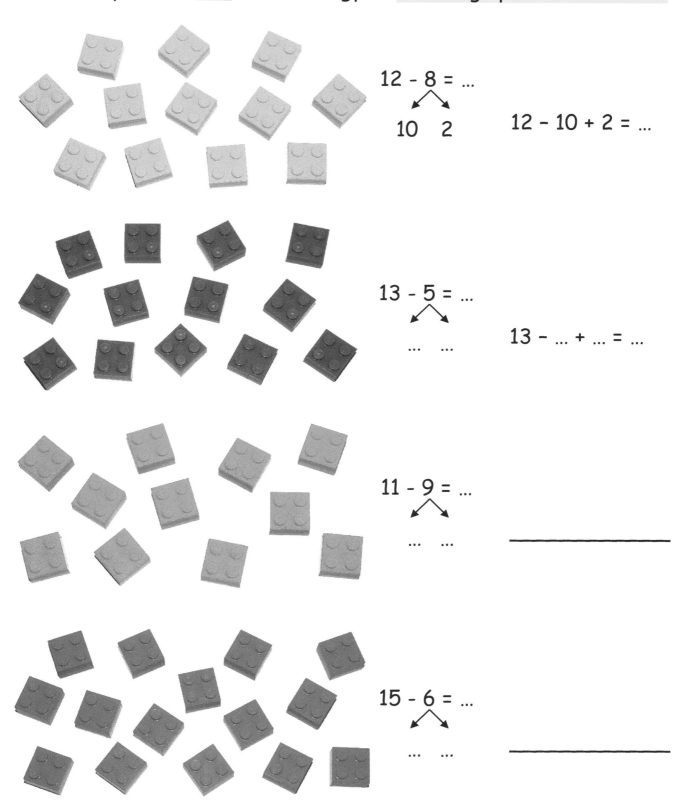

12 − 8 = …
   10   2

12 − 10 + 2 = …

13 − 5 = …
   …   …

13 − … + … = …

11 − 9 = …
   …   …

_____

15 − 6 = …
   …   …

_____

1. Number Sense Strategy. Fill in the missing numbers to complete each number sentence. Use the bricks. Circle the bricks to show the difference. You can subtract the leftover bricks in any order and quantity.

17 − 8 = 9

17 − ... − ... = 9

17 − ... − ... = 9

17 − ... − ... = 9

17 − ... − ... = 9

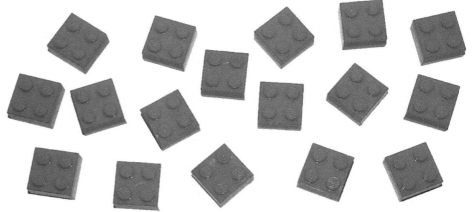

17 − ... − ... − ... = 9   17 − ... − ... − ... = 9

17 − ... − ... − ... = 9   17 − ... − ... − ... − ... = 9

17 − ... − ... − ... − ... = 9   17 − ... − ... − ... − ... − ... = 9

2. Write the number words and the value of the hundreds, tens, and ones.

458 Four hundred fifty-eight_____

458 4 hundreds_____   5 tens_____   8 ones_____

341 _____

341 ... _____ ... _____ ... _____

179 _____

179 ... _____ ... _____ ... _____

30

1. **Number Sense Strategy.** Write subtraction number sentences for each picture. Use the strategy of rounding up the subtrahend.

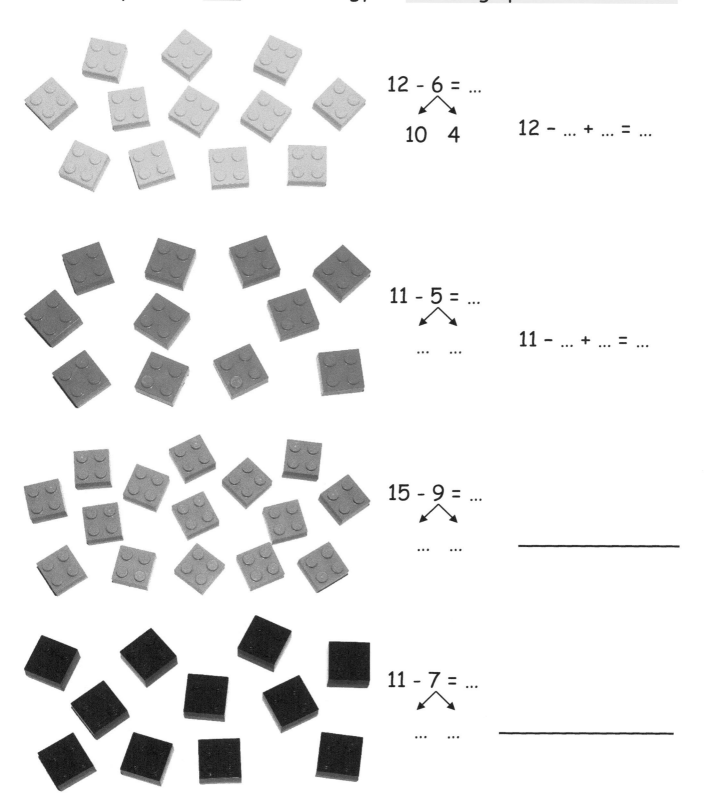

12 − 6 = ...
  10   4

12 − ... + ... = ...

11 − 5 = ...
  ...   ...

11 − ... + ... = ...

15 − 9 = ...
  ...   ...

_____

11 − 7 = ...
  ...   ...

_____

1. <u>Answer</u> the questions and <u>fill in</u> the missing letters.

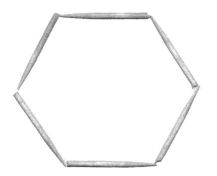 

<u>Take</u> 3 alike toothpicks and <u>break</u> them in the middle. <u>Connect</u> the ends of all the broken parts as the picture shows.

<u>Sketch</u> your shape in the box.

<u>How many sides</u> does it have?   … sides.

<u>How many angles</u> does it have?   … angles.

A closed shape with 6 sides and 6 angles is called a

… … … … … … …
8   5   24   1   7   15   14

(to solve this puzzle, look at the back of the book and find out what place each letter takes in the alphabet: A is 1ˢᵗ, Z is 26ᵗʰ).

2. <u>Look</u> at the picture, <u>answer</u> the question and <u>fill in</u> the numbers.

<u>How many more blue bricks</u> do I have?

… − … = … more blue bricks.

1. It's called a Hexagonal Honeycomb. <u>Why</u> did bees choose hexagons?_____

_____

_____

_____.

2. I have 20 bricks. My brother has 8 bricks. <u>How many more bricks</u> do I have?

I have	Brother has	How many more
...	...	? ..._____

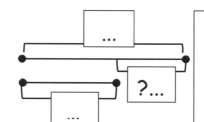

Answer:_____

3. <u>Find</u> 3 triangles in the shape and <u>write</u> down the letters, for example, ACD. The order of the letters does not matter.

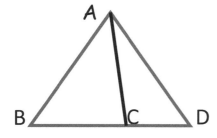

1. ... ...
2. ... ...
3. ... ...

4. <u>What day</u> is today? _____.

<u>What day</u> will it be in 9 days? _____.

1. Number Sense Strategy. Fill in the missing numbers to complete each number sentence. Use the bricks. Circle the bricks to show the difference. You can subtract the leftover bricks in any order and quantity.

18 − 9 = 9

18 − 6 − 3 = 9

18 − … − … = 9

18 − … − … = 9

18 − … − … = 9

18 − … − … − … = 9   18 − … − … − … = 9

18 − … − … − … = 9   18 − … − … − … − … = 9

18 − … − … − … − … = 9   18 − … − … − … − … − … = 9

2. We bought 22 pounds of blueberries. We used 8 pounds for making a jam. How many pounds were left?

Bought	Used	Left
…	…	? …

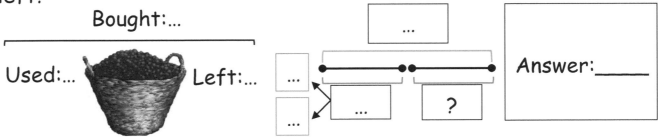

3. We had Tuesday the day before yesterday. What day of the week will it be after tomorrow? _____.

1. NSS. Find the value. Do you see a pattern in these problems?

| 14 – 7 + 6 = … |
| 14 – 6 + 7 = … |

| 15 – 5 + 6 = … |
| 15 – 6 + 5 = … |

| 16 – 7 + 9 = … |
| 16 – 9 + 7 = … |

| 13 – 8 + 4 = … |
| 13 – 4 + 8 = … |

I noticed that _____

_____

_____.

2. Write the numbers between:

367   …   …   370   134   …   …   137

895   …   …   898   471   …   …   474

546   …   …   549   621   …   …   624

3. Look at the picture below. Take … toothpicks to make the same shape. Take 2 toothpicks away so that you have 2 unequal squares. Draw your shape in the box.

Think, Smart Brainer!

1. I solved 13 problems. My friend solved 5 problems less. How many problems did my friend solve?

I solved	Friend solved	Friend solved
...	...	? ...

 I: ...

 My friend: ...

Answer:____

2. NSS. Find the value. Do you see a pattern in these problems?

13 − 9 + 8 = ...
13 − 8 + 9 = ...

14 − 7 + 9 = ...
14 − 9 + 7 = ...

15 − 6 + 9 = ...
15 − 9 + 6 = ...

16 − 8 + 4 = ...
16 − 4 + 8 = ...

I noticed that _____

_____

_____.

3. Let's guess the dimensions of your table, and then, measure it (use inches or centimeters).

How long is your table?

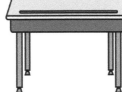

Guess: ... ____

True: ... ____

1. I helped Mom to bake ⟨24⟩ chocolate chip cookies and ⟨9⟩ sprinkle cookies. <u>Which cookie did we bake more of? How many more?</u>

Choco-late	Sprin-kle	How many more
…	…	? … _____

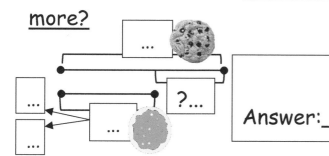

Answer:_____

2. Number Sense Strategy. <u>Fill in</u> the missing numbers to complete each number sentence. <u>Use</u> the bricks. <u>Circle</u> the bricks to show the difference. You can <u>subtract</u> the leftover bricks in any order and quantity.

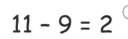

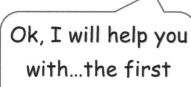

11 − 1 − 8 = 2   11 − … − … = 2

11 − … − … = 2   11 − … − … = 2

11 − … − … − … = 2   11 − … − … − … = 2

11 − … − … − … = 2   11 − … − … − … − … = 2

11 − … − … − … − … = 2 11 − … − … − … − … − … = 2

1. I jumped ⟦21⟧ times on the trampoline. My sister jumped ⟦9⟧ times. Who jumped more? How many more?

I jumped	Sister jumped	How many more
...	...	? ... ____

I: ...      Sis: ...

Answer:_____
_____

2. Number Sense Strategy. Fill in the missing numbers to complete each number sentence. Use the bricks. Circle the bricks to show the difference. You can subtract the leftover bricks in any order and quantity.

14 – 9 = 5

14 - ... - ... = 5

14 - ... - ... = 5

14 - ... - ... = 5

14 - ... - ... = 5

14 - ... - ... - ... = 5

14 - ... - ... - ... = 5

14 - ... - ... - ... = 5

14 - ... - ... - ... - ... = 5

14 - ... - ... - ... - ... = 5

14 - ... - ... - ... - ... - ... = 5

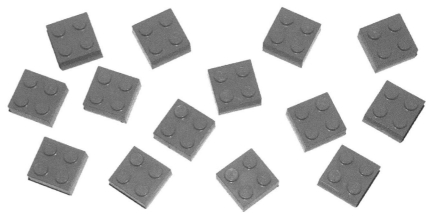

3. My brother and sister are ⟦11⟧ years old altogether. Brother is ⟦3⟧ years older than my sister. How old are they?
_____.

1. I need ☐23☐ invitation cards for my birthday party. I have made ☐7☐ cards. How many cards are left to be made? Fill in the table.

Need:...

Made:... Left:...

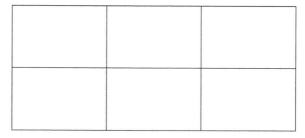

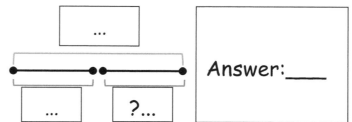

Answer:___

2. Find the value. Compare the fractions and circle >, <, or =.

$\frac{1}{2}$ > < = $\frac{3}{8}$

$\frac{2}{4}$ > < = $\frac{4}{8}$

$\frac{1}{2} + \frac{2}{2} = \frac{1+2}{2} = \frac{...}{...}$ > < = $\frac{2}{4} + \frac{2}{4} = \frac{...+...}{...} = \frac{...}{...}$

$\frac{5}{8} + \frac{2}{8} = \frac{...+...}{...} = \frac{...}{...}$ > < = $\frac{3}{4} + \frac{1}{4} = \frac{...+...}{...} = \frac{...}{...}$

3. Let's guess the dimensions of your table, and then, measure it (use inches or centimeters).

How wide is your table?   Guess: ...___   True: ...___

How high is your table?   Guess: ...___   True: ...___

1. I made ⟨15⟩ Valentine's cards for kids from my class and ⟨9⟩ less cards for kids from the swim team. How many cards will I make for kids from the swim team?

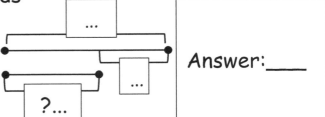

Answer:____

2. Number Sense Strategy. Fill in the missing numbers to complete each number sentence. Use the bricks. Circle the bricks to show the difference. You can subtract the leftover bricks in any order and quantity.

17 − 9 = 8

17 − ... − ... = 8

17 − ... − ... = 8

17 − ... − ... = 8

17 − ... − ... = 8

17 − ... − ... − ... = 8

17 − ... − ... − ... = 8

17 − ... − ... − ... − ... = 8

17 − ... − ... − ... = 8

17 − ... − ... − ... − ... = 8

17 − ... − ... − ... − ... − ... = 8

3. My sister is ⟨5⟩ years older than me. The sum of our ages equals ⟨17⟩ years. How old are we? _____

_____.

1. Last weekend I was fishing on a farm. I caught 15 fish, 6 of them were small. We put them back. How many fish did I bring home? Draw the diagram.

Caught	Let back	Brought home
...	...	? ...

Caught:...
Let back:...   Brought:...

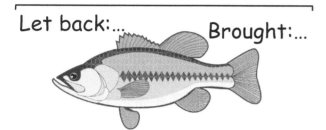

Answer:____

2. Number Sense Strategy. Fill in the missing numbers to complete each number sentence. Use the bricks. Circle the bricks to show the difference. You can subtract the leftover bricks in any order and quantity.

13 – 9 = 4

13 - 3 - 6 = 4

13 - ... - ... = 4

13 - ... - ... - ... = 4

13 - ... - ... - ... = 4

13 - ... - ... - ... - ... = 4

13 - ... - ... = 4

13 - ... - ... = 4

13 - ... - ... - ... = 4

13 - ... - ... - ... - ... = 4

13 - ... - ... - ... - ... - ... = 4

1. NSS. Find the value.

| 13 − 7 + 8 = … |
| 13 − 8 + 7 = … |

| 18 − 9 + 6 = … |
| 18 − 6 + 9 = … |

| 15 − 8 + 3 = … |
| 15 − 3 + 8 = … |

| 11 − 2 + 9 = … |
| 11 − 9 + 2 = … |

How do we calculate addition and subtraction in one equation?

_____

_____.

2. Our neighbor picked up 14 pounds of honey. He sold 9 pounds. How many pounds were left over? Fill in the table.

Picked:…
Sold:…   Left:…

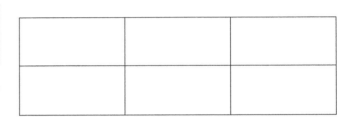

3. Find the value.

$\frac{1}{5} + \frac{1}{5} = \frac{…}{…}$

$\frac{1}{5} + \frac{4}{5} = \frac{…}{…} = …$

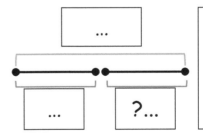

$\frac{1}{5}$

$\frac{2}{5} + \frac{3}{5} = \frac{…}{…} = …$

$\frac{3}{5} - \frac{1}{5} = \frac{…}{…}$

$\frac{5}{5} - \frac{4}{5} = \frac{…}{…}$

$\frac{4}{5} - \frac{1}{5} = \frac{…}{…}$

1. It takes us ⬚15 minutes to go to school. We have been driving for ⬚7 minutes. How much more time is left? Fill in the table and draw the diagram.

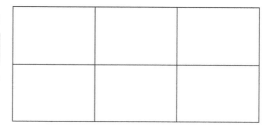

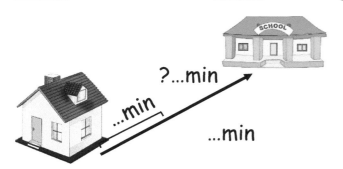

Answer:____

2. **Number Sense Strategy.** Fill in the missing numbers to complete each number sentence. Use the bricks. Circle the bricks to show the difference. You can subtract the leftover bricks in any order and quantity.

12 − 9 = 3

12 − 8 − 1 = 3   12 − ... − ... = 3

12 − ... − ... = 3   12 − ... − ... = 3

12 − ... − ... − ... = 3   12 − ... − ... − ... = 3

12 − ... − ... − ... = 3   12 − ... − ... − ... − ... = 3

12 − ... − ... − ... − ... = 3   12 − ... − ... − ... − ... − ... = 3

1. I scored |15| balls and missed |5| more balls than I scored. How many balls did I miss? Draw the diagram.

Hit	Missed	Missed
...	... _____	? ...

Answer:___

2. I played |15| soccer games and |5| basketball games this season. How many less basketball games did I play? Draw the diagram.

Soccer	Basket-ball	Basket-ball
...	...	? ... less

... _____

Answer:___

3. Fill in the missing numbers.

Hundreds	Tens	Ones	Number
6	2	7	...
9	5	3	...
2	1	8	...
5	8	6	...
1	3	5	...
...	...	...	721
...	...	...	467
...	...	...	953
...	...	...	348

1. <u>Answer</u> the questions, using a calendar.

How many days do you study in December? ... days.

<u>How many days</u> are weekends? ... days.

<u>How many days</u> are you on vacation? ...days.

How many days do you study in January? ... days.

<u>How many days</u> are weekends? ... days.

<u>How many days</u> are you on vacation? ...days.

How many days do you study in February? ... days.

<u>How many days</u> are weekends? ... days.

<u>What month</u> has the most weekends? _____

<u>What month</u> do you study the most? _____

<u>What month</u> is your favorite? _____

<u>Fill in</u> the column graph.

Winter

31 —
26 —
21 —
16 —
11 —
6 —
1 —

December   January   February

■ Days of vacation   ■ Weekends   ■ Days to study

1. <u>Fill in</u> the missing numbers in the Subtraction table. <u>Subtract</u> the number in a row from the number in a column. Some of them are done for you. If you see "-", it means you cannot find the difference (yet!). <u>Color</u> the boxes with the even numbers blue.

−	1	2	3	4	5	6	7	8	9	10	11	12	13	14	15	16	17	18
1	0	1	2	3	4	…	…	…	…	…	…	…	…	…	…	…	…	…
2	-	…	…	…	…	…	…	…	…	…	…	…	…	12	…	…	…	…
3	-	-	…	…	…	…	…	…	…	…	…	…	…	…	…	…	…	15
4	-	-	-	…	…	…	…	4	…	…	…	…	…	…	…	…	…	…
5	-	-	-	-	…	1	…	…	…	…	…	…	…	…	…	…	…	…
6	-	-	-	-	-	…	…	…	…	…	…	…	…	…	…	…	11	…
7	-	-	-	-	-	-	…	…	…	…	…	…	…	…	…	…	…	…
8	-	-	-	-	-	-	-	…	…	…	…	…	…	6	…	…	…	…
9	-	-	-	-	-	-	-	-	…	…	…	…	…	…	…	…	…	…

2. <u>Find</u> the value. <u>Compare</u> the fractions (<u>use</u> > or <).

$\frac{5}{5} - \frac{1}{5} = \frac{…}{…}$ … $\frac{4}{5} - \frac{1}{5} = \frac{…}{…}$ … $\frac{3}{5} - \frac{2}{5} = \frac{…}{…}$

$\frac{3}{5} + \frac{1}{5} - \frac{2}{5} = \frac{…}{…}$ … $\frac{4}{5} - \frac{2}{5} + \frac{3}{5} = \frac{…}{…}$ … $\frac{2}{5} + \frac{3}{5} - \frac{4}{5} = \frac{…}{…}$

$1/5$

1. <u>Color</u> the boxes with the black odd numbers green.

_	1	2	3	4	5	6	7	8	9	10	11	12	13	14	15	16	17	18
1	0	1	2	3	4	5	6	7	8	9	10	11	12	13	14	15	16	17
2	-	0	1	2	3	4	5	6	7	8	9	10	11	12	13	14	15	16
3	-	-	0	1	2	3	4	5	6	7	8	9	10	11	12	13	14	15
4	-	-	-	0	1	2	3	4	5	6	7	8	9	10	11	12	13	14
5	-	-	-	-	0	1	2	3	4	5	6	7	8	9	10	11	12	13
6	-	-	-	-	-	0	1	2	3	4	5	6	7	8	9	10	11	12
7	-	-	-	-	-	-	0	1	2	3	4	5	6	7	8	9	10	11
8	-	-	-	-	-	-	-	0	1	2	3	4	5	6	7	8	9	10
9	-	-	-	-	-	-	-	-	0	1	2	3	4	5	6	7	8	9

2. <u>Answer</u> the questions.

I have 2 sticks below. <u>How</u> can I make them equal?

_____

_____.

If I cut a stick, <u>should</u> it be the longest or the shortest stick?
_____.

<u>How many inches</u> should be cut? _____.

1. <u>Complete</u> each number sentence. <u>Try</u> to make 10's as it's easier to calculate them. You can <u>draw</u> the arrows to help you make 10's.

17 − 3 − 1 − 6 − 1 = …   13 − 8 − 2 − 1 − 1 = …

20 − 4 − 5 − 4 − 5 = …   20 − 5 − 7 − 4 − 2 = …

15 + 4 − 13 + 6 = …   14 − 7 + 10 − 9 = …

12 + 5 − 11 + 4 = …   11 − 7 + 15 − 12 = …

2. <u>Look</u> at the shape below. <u>Take</u> … toothpicks and <u>make</u> the same shape. <u>Take</u> 3 toothpicks <u>away</u> so that you will be left with 3 equal squares. <u>Draw</u> the shape you got in the box.

3. <u>Let's</u> guess the dimensions of your chair, and then, <u>measure</u> it (use inches or centimeters).

<u>How high</u> is your chair?   Guess: …____   True: …____

<u>How high</u> is your seat?   Guess: …____   True: …____

 Height of your chair

 Height of your seat

1. <u>Use</u> books, <u>Google,</u> or <u>ask</u> an adult to answer these questions.

The longest day of the year 2018 is June the 21ˢᵗ (called June or Summer Solstice). The shortest day of the year 2018 is December the 21ˢᵗ (called Winter Solstice). The days grow longer from December and grow shorter from June.

<u>When</u> are the days longer:

- in January or in February? _____
- in February or in March? _____
- in March or in May? _____

<u>When</u> are the days shorter:

- in July or in September? _____
- in October or in November? _____
- in December or in August? _____

There are 2 days of the year when the day almost equals the night. It's called an equinox. It comes around March the 20ᵗʰ or 21ˢᵗ and September the 22ⁿᵈ or 23ʳᵈ. On these days <u>how long</u> is:

- the night?   Around ... hours
- the day?   Around ... hours.

September the 22ⁿᵈ or 23ʳᵈ is the first day of Fall, and March the 20ᵗʰ or 21ˢᵗ is the first day of Spring.

<u>Do</u> you know any other days that tell us about the start of any season? _____

_____.

"I had 7 bricks. I added more and now I have 13 bricks. How many bricks did I add?" Don't you think, guys, this problem is too co-mpli-ca-ted... I don't get it 😐. I'm bad at math☹.

Wow! You do not understand a word, do you?! What will I do, then? I know, I should quit math! 🤮.

Calm down, guys ☺! Nobody is going to quit! Let's do it together.

I will help! "I had 7 bricks." Take 7 green bricks. It says, "I added more." If you do not know the number, name it "X" or put "?". I like to write "X": 7 + X. Then, you need to add as many bricks as you need to have 13 in all. Add blue bricks until you have 13 bricks altogether: 7 + X = 13. Do you have 13 bricks? Great! Now, count how many blue bricks you added.

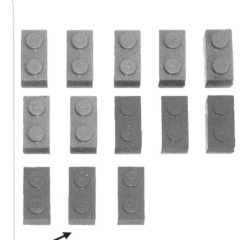

OK, I will try. I have 7 green bricks. Then, I add more blue bricks till I have 13 bricks altogether. Hm... Let me write it in the table.

Had	Added	Now have	Added
7	more	13	? ...

I make a number sentence for the problem: 7 green bricks + X equals 13.

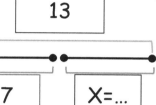

$7 + X = 13$

$X = 13 - 7$

$X = ...$

I know how to find X: 13 – 7 equals 6. Then, I write 6 in the table and diagram.

Aha, WHOLE minus PART...

I see. We added 6 blue bricks and we found out that 13 – 7 equals also 6 bricks.

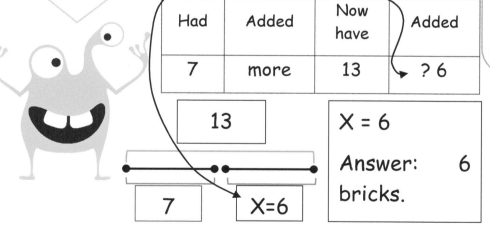

$X = 6$

Answer: 6 bricks.

1. <u>Compare</u> the number sentences in your mind, <u>use</u> the arrows to make 30's, and <u>fill in</u> ">", "<", or "=".

11 + 15 + 15 + 18 + 19 + 12   ...   14 + 19 + 13 + 11 + 16 + 17

1. I had 8 bricks. I added more and now I have 20 bricks. How many bricks did I add?

Had	Added	Have	Added
...	_____	...	? ...

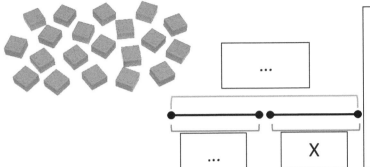

... + X = ...

X = ... − ...   X = ...

Answer:_____

2. NSS. Complete each number sentence. Make 10's with arrows.

18 − 6 − 2 − 4 − 4 = ...   15 − 1 − 7 − 3 − 1 = ...

18 − 3 − 8 − 2 − 5 = ...   20 − 5 − 7 − 4 − 3 = ...

14 + 5 − 17 + 9 = ...   13 − 8 + 12 − 6 = ...

16 + 3 − 14 + 6 = ...   19 − 3 + 4 − 15 = ...

3. I had bricks. I added 13 bricks and now I have 19. How many bricks did I have?

Had	Added	Have	Had
...	...	...	? ...

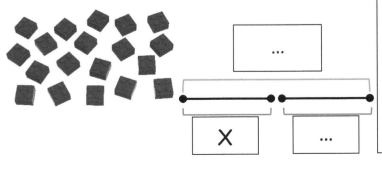

X + ... = ...

X = ... − ...   X = ...

Answer:_____

1. Number Sense Strategy. <u>Fill in</u> the missing numbers to complete each number sentence. <u>Use</u> the bricks. <u>Circle</u> the bricks to show the difference. You can <u>subtract</u> the leftover bricks in any order and quantity (e.g. = for example).

16 − 9 = 7   16 − … − … = 7

Did you notice a trick with 9? Look, 7 + … = 16: the ones in the sum is 1 less than the addend 7, right? Or 3 + … = 12. Yes, 3 + 9 = 12.

Aha, the second addend should be 9, as 9 needs 1 more to make 10: 7 − 1 = 6 (sum is 16), 3 − 1 = 2 (sum is 12).

16 − … − … = 7

16 − … − … = 7

16 − … − … − … = 7

16 − … − … − … = 7

16 − … − … − … − … = 7

16 − … − … − … = 7

16 − … − … − … − … = 7

16 − … − … − … − … − … = 7

2. <u>Answer</u> the questions.

<u>How many inches</u> is the stick? … in.

If I cut it into 3 equal parts, <u>what fraction</u> shows 1 part? $\frac{…}{…}$.

<u>How many inches</u> is 1 part? …in.

<u>How many inches</u> are 2 parts? …in.

<u>What fraction</u> shows 2 parts of the stick? $\frac{…}{…}$.

1. I got a set with 85 bricks. My little sister got a set with 45 bricks less. How many bricks were there in both sets?

I had	Sister had	Sister had	Both had
...	... ___	? ...	? ...

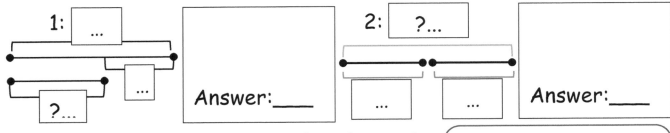

1: ...   Answer: ___

2: ? ...   .   Answer: ___

Write one number sentence for the problem: _____.

Aha, the first step is in parenthesis.

2. Complete each number sentence.

Fine☹. I calculate from the left to the right: 17-4=13, write 13; 13-10=3, write 3; 3+9=12, write 12; 12+5=17. Done!

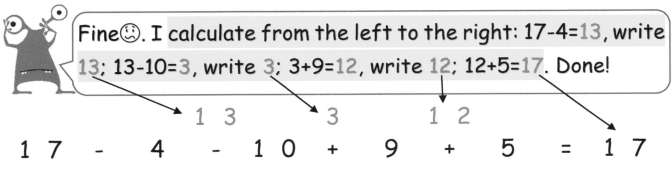

```
         1 3          3          1 2
17  -  4  -  1 0  +  9  +  5  =  1 7
```

```
        ...        ...        ...
15  -  8  -  3  +  1 4  -  7  =  ...
```

3. Let's guess the dimensions of your book, and then, measure it (use inches or centimeters).

How long is your book?   Guess: ... ___   True: ... ___

How wide is your book?   Guess: ... ___   True: ... ___

1. I had 14 candies. Yesterday I ate 5 of them. How many candies are left?

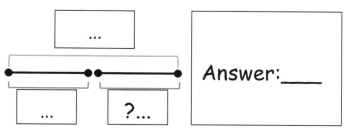

Answer:____

2. I had 14 candies. My brother had 5 candies less. How many candies did he have?

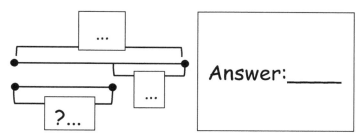

Answer:____

3. Look at the picture below. Take ... toothpicks and make 6 squares. Take away 2 toothpicks and make 7 equal squares with the remaining toothpicks. Draw the shape or shapes you got in the box.

1. Cut out 16 equal squares of 4 colors: 4 – orange squares, 4 – red squares, 4 – green squares, 4 – blue squares.

So, you have 16 squares of 4 colors. Draw 1 big square divided into 16 small squares.

Arrange the squares so that each row, column, and numeral diagonal has 4 different colors.

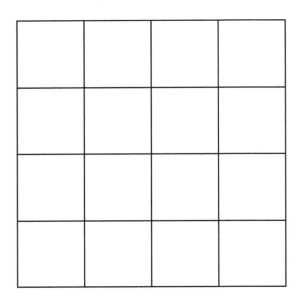 

2. Complete each number sentence.

12 + 6 - 9 + 7 + 3 = …

18 - 5 - 6 + 11 - 4 = …

12 + 3 - 9 - 2 + 7 = …

1. I have ☐13 red pens and ☐7 less black pens. <u>How many black pens</u> do I have? <u>Draw</u> the diagram.

Red	Black	Black
...	... _____	? ...

 ...

 ... _____

Answer:____

2. I played ☐12 soccer games and ☐4 basketball games this season. <u>How many less basketball games</u> did I play? <u>Draw</u> the diagram.

Soccer	Basket-ball	Basket-ball
...	...	? ... less

Answer:____

3. My little brother has ☐12 teeth. Last year he had only ☐7 teeth. <u>How many teeth</u> did he get this year?

Has	Had	Got this year
...	...	? ...

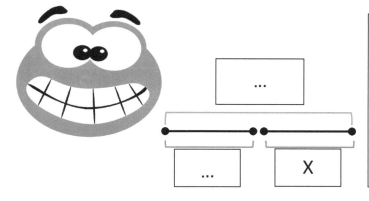

... + X = ...

X = ...

X = ...

Answer:_____

1. Number Sense Strategy. Fill in the numbers to complete each number sentence. Circle the bricks to show the difference. You can subtract the leftover bricks in any order and quantity.

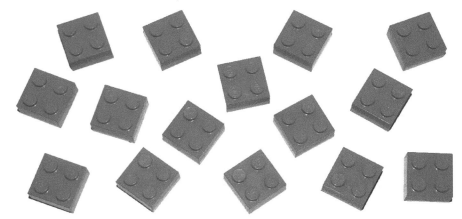

15 – 9 = 6

15 - ... - ... = 6

15 - ... - ... = 6

15 - ... - ... = 6

15 - ... - ... = 6

15 - ... - ... - ... = 6

15 - ... - ... - ... - ... = 6

15 - ... - ... - ... - ... = 6

15 - ... - ... - ... - ... - ... = 6

2. What bills and coins may make up $12.50? Fill in the missing coins and bills.

1 5-dollar bill, ... 1-dollar bill, 10 quarters, ... dimes, ... nickels, 50 pennies.

... 5-dollar bill, 4 1-dollar bill, ... quarters, ... dimes, 40 nickels, 30 pennies.

3. A half of a quart of milk costs $4. How much does a quart of milk cost?

A half	Halves in a quart	Quart of milk
...	...	? ...

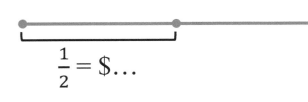

$\frac{1}{2} = \$...$

Answer: _____.

1. My sister is ⌐6¬. If I add her age to mine, the sum will be ⌐14¬. How old am I now?

Sister is	Add and I will be	I am now
...	...	? ...

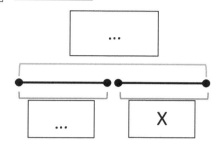

... + X = ...

X = ... − ...   X = ...

Answer:_____.

2. Find X making calculations in your mind. X = WHOLE − PART2.

5 + X = 12   8 + X = 11   7 + X = 13

X = ...   X = ...   X = ...

8 + X = 16   9 + X = 13   6 + X = 15

X = ...   X = ...   X = ...

9 + X = 11   5 + X = 13   8 + X = 17

X = ...   X = ...   X = ...

3. Look at the picture below and answer the questions.

How many sticks are there? _____.

How many intervals are there? _____.

How many steps are there in all?

_____

_____.

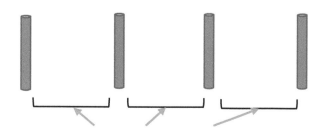

Intervals = 10 steps

1. There were 7 boys in our Gym team. Last month more kids joined the team. Now we have 12 boys. How many boys joined the team?

Were	Joined	Now	Joined
...	_____	...	? ...

... + X = ...

X = ...

X = ...

Answer: _____.

2. It's 2:45pm now. Draw the minute and hour hands red. I go to sleep at 8:25pm, draw the minute and hour hands black.

How much time is left until I go to sleep?

_____.

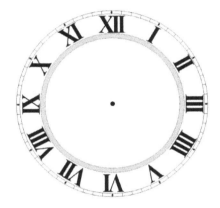

3. Grandma picked up 19 pounds of raspberries in the garden. After she gave 7 pounds to us, 6 pounds were left. How many pounds did she use for jam?

Picked up	Sold out	Were left	Jam
...	...	...	? ...

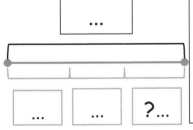

Answer: _____.

1. <u>Find</u> a number behind a creature, making calculations in your mind.

17 - 🐿 = 9   13 - 🐱 = 5   15 - 🐰 = 6

🐿 = ...   🐱 = ...   🐰 = ...

13 - 🐌 = 8   11 - 🐊 = 2   16 - 🐉 = 9

🐌 = ...   🐊 = ...   🐉 = ...

12 - 🐒 = 6   14 - 🐜 = 5   11 - 🐢 = 8

🐒 = ...   🐜 = ...   🐢 = ...

2. <u>How many angles</u> do you see here? _____.

<u>How many</u> are obtuse angles? _____.

<u>How many</u> are acute angles? _____.

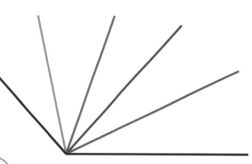

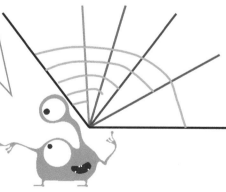

Look, I explore angles in the black-black lines and mark them orange: 1 + 1 + 1 + 1 + 1 = 5. Then, you will take blue-black lines, then, red-black lines, then, green-black lines, at last, gold-black lines. Done!

I will take blue-black lines: blue-red + blue-green + blue-gold + blue-black. 1+1+1+1=4.

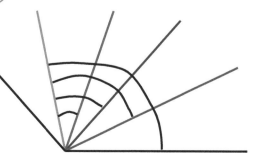

1. I found 29 4-stud bricks and 11 more 6-stud bricks. How many bricks did I find in all?

4-stud	6-stud	6-stud	In all
...	... ___	? ...	? ...

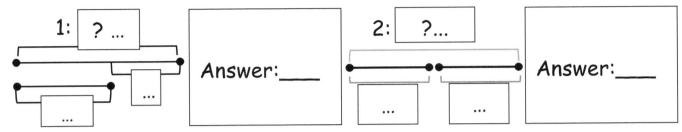

Write one number sentence for the problem:

_____.

2. I poured 10 halves of a cup. How many cups are now in the pan?

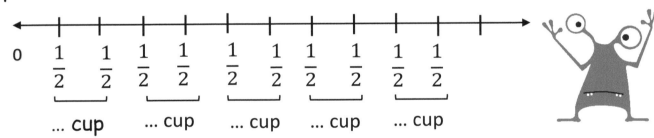

1 cup = ... halves of a cup. So, 1 cup = $\frac{...}{...} + \frac{...}{...}$ =

Answer: _____.

3. Cross out the picture which is different.

1. The sum of ⬚3 numbers is ⬚20. The sum of the first and second numbers is ⬚12. The difference of the third and second numbers is ⬚1. Find the numbers.

_____.

2. Insert the missing numbers hiding behind my pets.

... + ... + ... + ... = 44

... + ... + ... = 31

... + ... − ... = 16

3. Complete the number sentence.

1 7 − ... 4 − ... 1 0 + ... 9 + 5 = ...

4. Write in the missing numbers to balance my toys.

4 owls = ... frogs   3 frogs = ? ... owl(s)

1. For December let's make a project "Daytime" ☺.

You need to check the day length every day and write the hours and minutes of sunrise and sunset in the table. You can use websites Solartopo.com, Timeanddate.com, Aa.usno.navy.mil or you can use Google.com. Length – is the difference between the sunset and the sun rise. Dif. – is the difference between the lengths of 2 days. It's "+" if the daytime goes up, and it's "-" if the daytime goes down. The Dif. Is unknown for the 1st of December.

Date	Day length			Dif.	Date	Day length			Dif.
	Srise	Sset	Length			Srise	Sset	Length	
1					16				
2					17				
3					18				
4					19				
5					20				
6					21				
7					22				
8					23				
9					24				
10					25				
11					26				
12					27				
13					28				
14					29				
15					30				
					31				

Add all the differences for the day length for the whole month: _____.

1. <u>Make</u> two Line graphs, one for Sunrise and one for Sunset time in December in your area according to your table.

A line Graph shows how the data changes continuously over time.

<u>Use</u> dots to show the time and line segments to connect the dots and show how the time changes.

The dots and line segments for Sunrise are blue and the dots and line segments for Sunset are red.

Sunrise in December

Sunrise Time

	1	2	3	4	5	6	7	8	9	10	11	12	13	14	15	16	17	18	19	20	21	22	23	24	25	26	27	28	29	30	31
Srise																															

Sunset in December

Sunset Time

	1	2	3	4	5	6	7	8	9	10	11	12	13	14	15	16	17	18	19	20	21	22	23	24	25	26	27	28	29	30	31
Sset																															

2. ③ Peaches and ③ oranges are in balance with ② oranges and ④ peaches. Which fruit is heavier: a peach or an orange? _____

_____

_____ .

1. <u>Compare</u> the fractions. <u>Use</u> "=", ">" or "<" and the shape below.

$\frac{3}{4}$ ... $\frac{3}{8}$ $\frac{2}{8}$ ... $\frac{2}{12}$

$\frac{1}{4}$ ... $\frac{2}{8}$ $\frac{2}{4}$ ... $\frac{4}{12}$

$\frac{9}{12}$ ... $\frac{3}{4}$ $\frac{2}{8}$ ... $\frac{3}{12}$

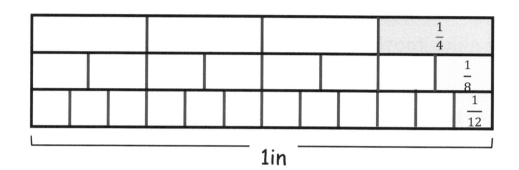

2. My sister, brother, and I had 9 candies. Sister gave 4 candies to Brother, and Brother gave me 2 candies, then, we all had the same amount of candies. <u>Look</u> at the diagram, <u>write in</u> the missing numbers for X, and <u>complete</u> the number sentences for each of us. <u>How many candies</u> did each have in the beginning? <u>How many candies</u> did each have in the end?

In the beginning: Sister: _____, Brother: _____, I: _____.

In the end: Sister: _____, Brother: _____, I: _____.

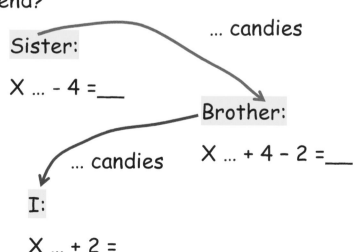

1. <u>Answer</u> the questions and <u>fill in</u> the missing letters.

<u>Take</u> 4 alike toothpicks. <u>Bend (or break)</u> each toothpick in the middle.

<u>Connect</u> the ends of all toothpicks as you see in the picture.

<u>Sketch</u> your shape in the box.

<u>How many sides</u> does it have?   … sides.

<u>How many angles</u> does it have?   … angles.

A shape with 8 straight sides and 8 angles is called

…   …   …   …   …   …   …
15   3   20   1   7   15   14

(to solve this puzzle, <u>look</u> at the back of the book and <u>find</u> out what place each letter takes in the alphabet: A is 1st, Z is 26th).

If I divide my octagon into 8 equal parts, <u>what fraction</u> will show 1 part? $\frac{…}{…}$.

<u>How many parts</u> will make a half of an octagon?   … parts.

<u>What kind of animal (arachnid)</u> has 8 legs? <u>Write</u> 3 examples:

_____.

1. <u>Fill in</u> the numbers instead of the letters. 1 letter means 1 digit (tens or ones).

A9 − B = 24   6C + D = 77   9E + F = G00

…9 − … = 24   6… + … = 77   9… + … = …00

2. I put ⬚ red balls on the Christmas tree. My sister put ⬚ red balls. <u>Who</u> put more red balls? <u>How many more?</u>

I put	Sister put	How many more
…	…	? … _____

I:…

Sis:…

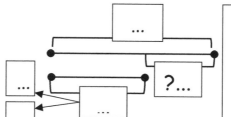

Answer:____

3. <u>Complete</u> the number sentence.

1 1 - 4 + 1 2 - 7 + 3 = …

4. <u>How many square units</u> are there in each shape? 1 cube equals 1 square unit.

… square units

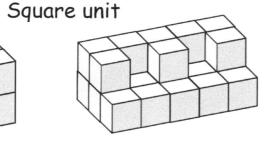
… square units

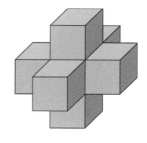
… square units

Do you know the circumference is the length around the circle, and the circle is the closed shape? Every dot on the circumference is equally distant from the center.

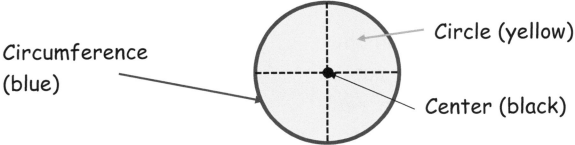

Circumference (blue)

Circle (yellow)

Center (black)

1. <u>Find</u> 8 triangles in the shape and <u>write</u> the letters. For example, AFC. The order of the letters does not matter.

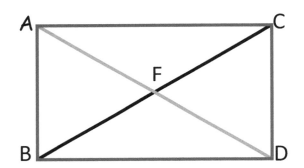

1. A F C   2. ... ...

3. ... ...   4. ... ...

5. ... ...   6. ... ...

7. ... ...   8. ... ...

2. Look at the picture below. <u>Take</u> ... toothpicks and make the same shape. <u>Move</u> 3 toothpicks to get 3 triangles. <u>Draw</u> your shape in the box.

Do you know the distance to the center of the Earth is 3,959 mi (6,371km)? It's the same as flying from New York City, USA to Berlin, Germany by plane (7:45 flight long).

1. How many parts are in this bar?... parts.

   What part is ⬚1 red block?   .../...

   If it's ⬚8 cm long, how long is $\frac{1}{4}$?   ... cm.

   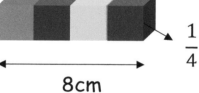

   How long is $\frac{2}{4}$?   _____   ... cm.

   How long is $\frac{3}{4}$?   _____   ... cm.

2. Grandma picked up peaches in the garden. After having cooked jam with 9 pounds of peaches, ⬚5 pounds were left over. How many pounds did she pick up?

Picked up	Cooked	Were left	Picked up
___	...	...	? ...

   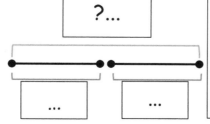

   ... + ... = X

   X = ...   X = ...

   Answer: _____

Do you know the deepest drill hole in the Earth is: Kola Superdeep Borehole on the Kola peninsula in Russia – 7.619 mi (12.262km).

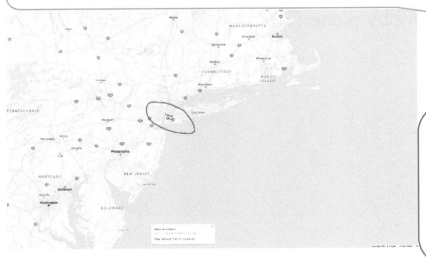

Wow! To reach the core = to cross the ocean, but the deepest drill hole is a 10-minute drive.

1. How many parts are in this bar?

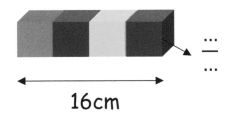

… parts.

What part is 1 yellow block? $\frac{...}{...}$.

16cm

If it's 16 cm long, how long is $\frac{1}{4}$? … cm.

How long is $\frac{2}{4}$? _____ … cm.

How long is $\frac{3}{4}$? _____ … cm.

2. Complete each number sentence.

1 5 - 8 - 3 + 1 4 - 7 = …

1 9 - 1 3 + 8 - 1 1 + 1 5 = …

I have 3 red bricks, 2 green bricks, and as many brown bricks as red and green bricks together. How many brown bricks do I have?

I have "as many brown bricks as red and green bricks together". So, I need to add them to find out how many of them together I have.

Red	Green	Brown	Brown
3	2	Red + Green	?...

?...

3   2

3 + 2 = …

Answer: I have … brown bricks.

Got it!

1. <u>Draw</u> the missing shape in the box and <u>write</u> the shapes' names.

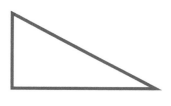  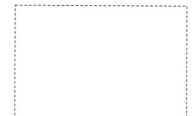 

_____   _____   _____   _____

I shared 7 bricks with my sister and 9 bricks with my brother. I didn't have any leftover. How many bricks did I have before sharing? I don't get it☹.

Calm down, we can always ask Smarty, he will be more than happy to "teach" us☹.

Bricks which I shared were the only bricks which I had. So, I need to add up what I shared to find out how many I had before sharing.

Had	Shared with sister	Shared with brother	Had before sharing
bricks	7	9	?...

?...

| 7 | 9 |

7 + 9 = …

Answer: I had … bricks.

1. Mom gave 12 candies to me and my sister. I divided them, but my sister said the parts were not equal. I had to give her 2 more candies, and then, our parts were equal. How many candies did we have in the beginning?

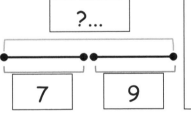

I'd had …   She'd had …

I have …   She has …

1. We bought 9 lb. of potatoes, 1 lb. of carrots, 2 lb. of cucumbers, 4 lb. of tomatoes and 5 lb. of apples. I helped to carry 10 lb. to the car. What vegetables and/or fruits could I carry? Draw the diagrams.

Potatoes	Carrots	Cucumbers	Tomatoes	Apples	10 lb.	10 lb.
					?	?
...	...	...	...	...		

_____

_____

1: [Answer:___   ]   2: [Answer:___   ]

2. Draw the minute and hour hands red for 10:50am. I have a 30-min recess, a 30-min chess class, a 40-min reading, and a 40-min art class. When will school be over? ...:... _____.

Draw the minute and hour hands black.

3. I put 8 apples and 7 pears on the plate. We ate 9 fruits. How many fruits were left?

Put	In all	Ate	Left
...	?...	...	?...
...			

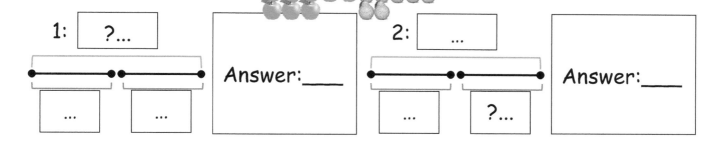

1. <u>Answer</u> the questions. <u>Draw</u> 3 women at the table.

There are 3 women sitting at the table: 2 mothers, 2 daughters, and a Grandmother with a granddaughter. <u>Who</u> are they?

Are you okay? I've counted 6 women at the table!!!

<u>Who</u> are they? They are: _____

_____.

2. <u>Answer</u> the questions and <u>fill in</u> the missing numbers.

<u>How many parts</u> are in this bar?   … parts.

<u>What fraction</u> shows 1 block? $\dfrac{…}{…}$.

If it's 80 cm long, <u>how long</u> is $\dfrac{1}{8}$?   … cm.

<u>How</u> did you find it?   _____.

<u>How long</u> is $\dfrac{3}{8}$?   … cm.

<u>How long</u> is $\dfrac{6}{8}$?   … cm.

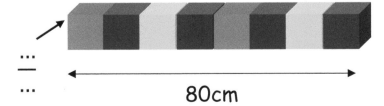

$\dfrac{…}{…}$ 80cm

8 + 5
/ \
2 3

	tens	ones
		...
		8
+		5
	...	...

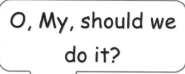 O, My, should we do it?

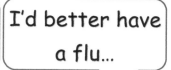 I'd better have a flu...

 I think I'm sick...

 No, why? It's so simple, guys! Well. Maybe not so simple, but still, you need to remember 2 things. First, how to make numbers up to ten (as you know, tens are very easy to count). The easiest way is to take the bigger number (8>5). We will make 10 by adding to 8.

Aha, the easiest way to make 10 out of 8 is to add 2:   8 + 2 = 10!

But, look, we are left with 3. Should we add 3 to 10?   10 + 3 = 13.   Yes.

1. My friends and I have played chess games. Each has played an equal number of games. How many games have we played in all? _____. How many games has each of us played? _____. Use the diagram to help you solve the problem.

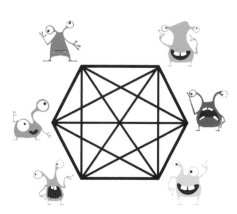

I want to explain calculations in columns! 8 + 5 is the same as 8 + 2 + 3 = 13. I write 3 in the ones column and write 1 above in the tens column.

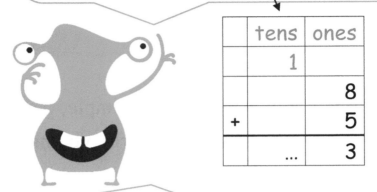

I don't see any more tens, so, I rewrite 1 in the total. It's the same as you've said, Smarty: 8 + 5 = 8 + 2 + 3 = 10 + 3. But I prefer calculations in columns.

1. I want to buy a book for $5.50, a set of gel pens for $9.50. I found in my piggy bank: 2 5-dollar bills, 4 1-dollar bills, 10 quarters, 15 dimes, 20 nickels, and 100 pennies.

How much did I have in my piggy bank?

_____.

How much will be left after my purchases?

_____

_____.

8 + 5

2 3

	tens	ones
	1	
		8
+		5
	1	3

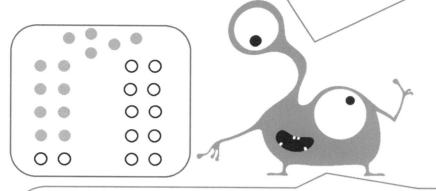

I want to explain the addition with the dots!

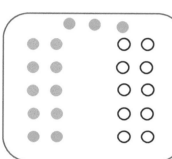

You have 8 yellow dots and 5 green dots. To add them together we need to count how many more dots are needed to be added to yellow dots to make up 10. 2 more dots! Yellow dots need 2 more dots to be a ten.   8 + 2 = 10

Aha... Now we have ten dots (8 + 2) and 3 more green dots are left.

5 − 2 = 3

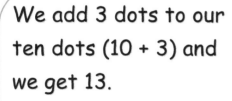

We add 3 dots to our ten dots (10 + 3) and we get 13.

10 + 3 = 13 or

8 + 2 + 3 = 13

8 + 5 = 13

1. This month we have had ⑦ days below freezing and ⑤ days around freezing. How many freezing days have we had this month?

Below freezing	Around freezing	Freezing days
...	...	? ...

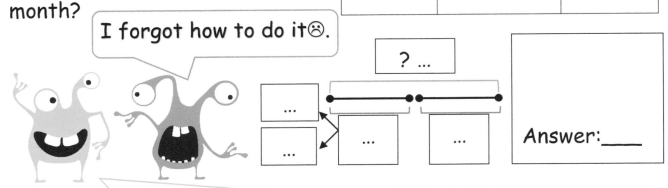

I forgot how to do it☹.

Answer:___

Take the bigger number. It's 7! Find out what turns 7 into ten. It is 7 + 3. You need to add 3 to 7 to get 10.
Look for a word family for 5 with 3. It's 3 + 2. So,
    7 + 3 + 2 = 10 + 2 = 12 (1 tens 2 ones).

2. <u>Find</u> X by mental math. Remember: X(Part1) = Whole − Part2.

X + 5 = 11   X + 7 = 14   X + 6 = 15

X = ...   X = ...   X = ...

X + 8 = 12   X + 9 = 14   X + 5 = 13

X = ...   X = ...   X = ...

X + 9 = 11   X + 7 = 12   X + 8 = 11

X = ...   X = ...   X = ...

3. <u>Continue</u> the pattern: 100, 95, 85, ..., ..., ..., ..., ...

1. Draw 3 unequal quadrilaterals with a ruler. Cut each one into 4 triangles. How many triangles did you get in each quadrilateral? … triangles, … triangles, … triangles.

Draw the shapes of your biggest and smallest triangles in the box.

The biggest triangle: The smallest triangle:

2. I had 15 chocolate cookies and 8 butter cookies. I gave 6 chocolate cookies to my sister and she gave me 7 butter cookies. How many chocolate cookies do I have now? How many butter cookies do I have now?

I - ch.	I - but.	I → Sis	Sis → Me	I - ch.	I - but.
…	…	…	…	?...	?...

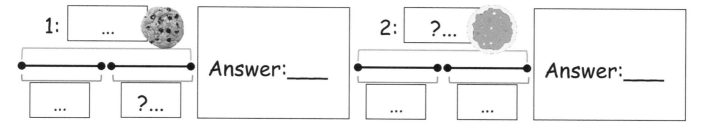

1: [… …] Answer:___

2: [?... …] Answer:___

1. <u>Draw</u> the minute and hour hands red for 10am .

How much time is left until 3:55 pm ?

... hours ... minutes

<u>Draw</u> the minute and hour hands black.

2. I have 7 chocolate cookies and my sister has 6 chocolate cookies. My brother has as many chocolate cookies as my sister and me. <u>How many chocolate cookies</u> does he have? <u>Fill in</u> the table.

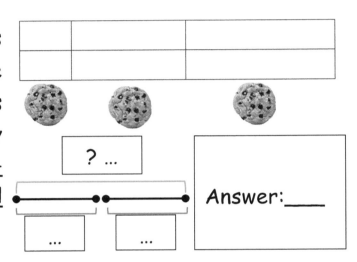

3. <u>Draw</u> the missing shape in the box and <u>write in</u> their names.

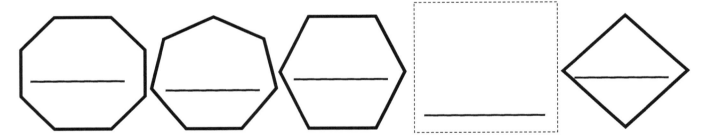

4. <u>Find</u> 3 triangles in the shape and <u>write</u> the letters, for example, CBD, in the box. The order of the letters does not matter.

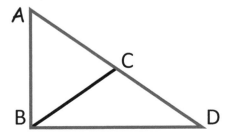

1. ... ...
2. ... ...
3. ... ...

1. Number Sense Strategy. **Complete** each number sentence and fill in the missing numbers. The first one is done for you.

   8 + 5   7 + 8   9 + 3
    ↙ ↘   ↙ ↘   ↙ ↘
   2   3   ... ...

	tens	ones
	1	
		8
+		5
	1	3

	tens	ones
	...	
		...
+		...
	...	...

	tens	ones
	...	
		...
+		...
	...	...

   8 + 5 = 8 + 2 + 3 =   7 + 8 = ... + ... + ... =   9 + 3 = ... + ... + ... =
   10 + 3 = 13   ... + ... = ... + ... = ...

2. **Explain** how we subtract using the pictures below.

18 − 9 = 18 − ... − ... = ...
      ↙ ↘
     8   ...

18

I would build a fortress, a castle, and an ARMY with these sticks!

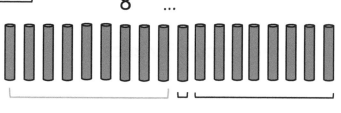

? ...   1   8

3. **How many** do you need to make ⟦1 whole⟧? Write in the missing fraction.

   $\frac{1}{2} + \frac{...}{...}$ $\frac{3}{4} + \frac{...}{...}$ $\frac{5}{6} + \frac{...}{...}$ $\frac{2}{3} + \frac{...}{...}$ $\frac{9}{10} + \frac{...}{...}$ $\frac{4}{5} + \frac{...}{...}$ $\frac{7}{8} + \frac{...}{...}$

1. <u>Sketch</u> a picture of yourself in the box. <u>Take</u> a measuring tape and <u>measure</u> yourself. <u>Write down</u> the measurements:

- Standing height   …_____

- Sitting height   …_____

- Head length   …_____

- Neck circumference   …_____

- Right shoulder   …_____

- Left shoulder   …_____

- Waist breadth   …_____

- Shoulder to elbow   …_____

- Forearm hand length   …_____

- Foot length   …_____

- Knee circumference   …_____

I am _____.

Chest breadth (under the armpits) with exhalation: …____

Chest breadth (under the armpits) with inhalation: …____

<u>Did</u> you notice any difference? <u>Why</u> is the chest breadth different with inhalation and exhalation?

_____.

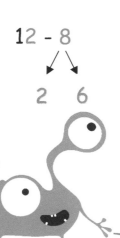

12 − 8 =
12 − 2 − 6 =
10 − 6 = 4

We've already learned that! I'm 🤢.

No, no! That's fun! Real fun! Maybe not so fun, but I DO love these problems. Just remember: We always subtract the ones out of the ones and the tens out of the tens ! You see that 12 has the ones digit 2 and the tens digit 1. You cannot subtract the ones digit 8 out of the ones digit 2. But you can represent 8 as 2 + 6. Then, 12 − 2 − 6 = 10 − 6 = 4. So, we make 10 out of 12 by subtracting 2.

1. I've cut the 10-inch log into 2 equal pieces. How long is 1 part?

How many cuts did I make? ....

Log	Equal pieces	Length of 1 part
...	...	? ...

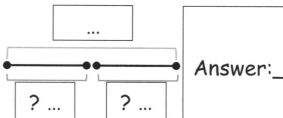

Answer:____

Guys, I will explain calculations in columns. I want to play football, so, let's do it fast! So, you need one more row above your problem. Well, you cannot subtract 8 out of 2. But you can borrow 10 out of tens. As 1 tens equals 10.

tens	ones
...	12
1̶	2̶
-	8
	4

Sure, what is 12? 12 = 10 + 2. Let's borrow ten and write 12 above 2, cross out 1 and 2 (they really get on my nerves – ha-ha). Can we subtract 8 out of 12? Easily! 12 – 8 = 4, we write 4 in the ones.

YOU are really getting on my nerves! Why do YOU always in-ter-rupt ME?! Gggrrr... Ok, let's finish. MY FOOTBALL!!!

When we borrowed 1 from tens, we have no more tens and write 0 above 1 which we have ALREADY crossed out. So, 0 minus ... nothing equals zero.

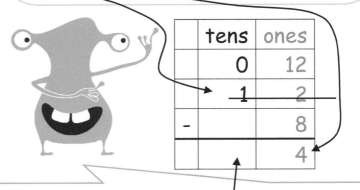

Ah, we do not write zeros in the difference before a digit! We leave an empty space. The answer is 4.

1. Number sense Strategy. Complete each number sentence and write in the missing numbers. The first one is done for you.

   11 - 2   11 - 4   11 - 7
   ↙ ↘   ↙ ↘   ↙ ↘
   1   1   ... ...

tens	ones
0	11
~~1~~	~~1~~
-	2
	9

tens	ones
...	...
...	...
-	...
	...

tens	ones
...	...
...	...
-	...
	...

   11 - 2 = 11 - 1 - 1 =   11 - 4 = ... - ... - ... =   11 - 7 = ... - ... - ... =
   10 - 1 = 9   ... - ... = ... - ... = ...

2. Find the value. Remember: 1 hour = 60 minutes.

   ```
     1 h 1 5 m          2 h 5 0 m          3 h 3 0 m
   + 1 h 1 5 m        + 4 h 2 0 m        + 1 h 3 5 m
   ─────────────      ─────────────      ─────────────
    ... h ... ... m    ... h ... ... m    ... h ... ... m
   ```

3. Draw the minute and hour hands red for midnight.

   How much time is left until 25 past 9 am?

   ... hours ... minutes

   Draw the minute and hour hands black.

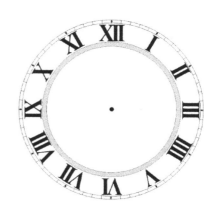

1. <u>Find</u> X by mental math. Remember: X = Whole = Part1 + Part2.

X - 7 = 6	X - 5 = 9	X - 9 = 2
X = ...	X = ...	X = ...
X - 4 = 8	X - 8 = 9	X - 6 = 9
X = ...	X = ...	X = ...
X - 9 = 9	X - 8 = 8	X - 7 = 8
X = ...	X = ...	X = ...

2. I turned 8 years old in 2017. <u>When</u> was I born? <u>Draw</u> the diagram.

   2017: 8yo

   Born: ? ...

Turned	Year	Was born
...	...	? ...

Answer:___

3. There are 15 cards in my album. There are 10 cards less in my brother's album. <u>How many cards</u> do we have altogether?

My	Brother's	Brother's	Altogether
...	...___	?...	?...

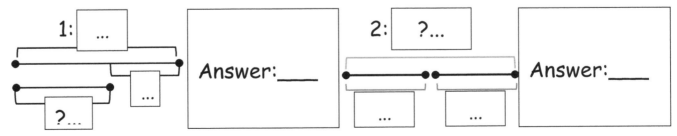

1: ...   ?...

Answer:___

2: ?...

Answer:___

<u>Circle</u> the correct number sentence: 15 + (15+10) or (15+10)+15
15 + (15-10) or (15-10)+15.

12 - 8
  ↙ ↘
  2   6

tens	ones
0	12
1	2
-	8
	4

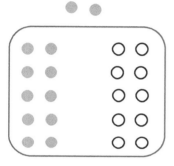

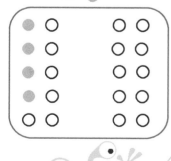

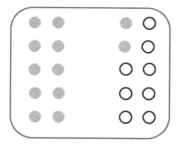

I have 12 ugly disgusting brown dots and I need to subtract 8 out of them. What is 12? 12 = 10 + 2. I like to count by tens, it's easy and simple. So, I need to get 10 out of 12. How? Subtract 2.

So, 12 – 2 = 10.

Look, we took away 2 dots and we have 10 dots. As 8 = 2 + 6, and we have already subtracted 2 dots, 6 more dots are left.

Then, 10 – 6 = 4.

Guys, I want to repeat what we did:

12 – 2 = 10

(8 = 2 + 6)

10 – 6 = 4

Or 12 – 2 – 6 = 4

And I WANT to go play football!!! Anyone with me?...

1. Number Sense Strategy. Complete each number sentence and write in the missing numbers.

11 - 9	12 - 3	12 - 8
... ...	... ...	... ...

	tens	ones
	...	...
	...	...
-		...
		...

	tens	ones
	...	...
	...	...
-		...
		...

	tens	ones
	...	...
	...	...
-		...
		...

11 - 9 = ... - ... - ... = ... - ... = ...

12 - 3 = ... - ... - ... = ... - ... = ...

12 - 8 = ... - ... - ... = ... - ... = ...

2. Draw the minute and hour hands red for 2pm. How much time is left until 15 past 11 pm?

... hours ... minutes

Draw the minute and hour hands black.

3. Measure 1 cup, 1 pint, 1 quarter, and 1 gallon of water. Find the connections between these measurements:

1 pint = ... cups   1 quart = ... cups   1 gallon = ... cups

1 quart = ... pints   1 gallon = ... pints   1gallon = ... quarts

1 gallon = ... quarts = ... pints = ... cups

1. <u>Choose</u> 2 weeks for weather observation. <u>Check</u> the sky at 11am and 6pm and <u>fill in</u> the table.

The sky may be:

Clear   Cloudy: $\frac{1}{4}$ Cloudy: $\frac{1}{2}$ Cloudy: $\frac{3}{4}$ Cloudy

Precipitation may be:

Rain   Thunder with lightning   Snow

The wind may be:

North   East   South   West

Day	Sky		Wind		t°		Precipitation	
	11am	6pm	11am	6pm	11am	6pm	11am	6pm

1. <u>Answer</u> the questions.

<u>How many clear days</u> did you check? _____.

<u>How many cloudy days</u> did you check? _____.

<u>What kind of cloudy days</u> did you observe the most? _____.

<u>How many windy days</u> were there? _____.

<u>How many rainy days</u> were there? _____.

<u>How many days</u> had thunderstorms? _____.

<u>Circle</u> the correct word: more/less and <u>answer</u> the questions.

<u>How many more/less clear days than rainy days</u> did you have?
_____.

<u>How many more/less windy days than rainy days</u> did you have?
_____.

<u>Draw</u> the columns for different kinds of days.

Project Weather: Days

	Rainy days	Snowy days	Clear days	Cloudy days	Windy days
14					
12					
10					
8					
6					
4					
2					
0					

1. I've done a lot of art projects on Christmas vacation: 11 coloring pages, 34 oil paintings, 21 pencil drawings, and 9 Christmas 3D structures. How many art projects have I done in all?

Coloring	Paintings	Drawings	3D structures	In all
...	...	...	...	?...

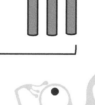

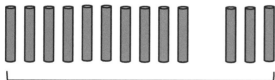

Answer:_____.

2. Explain how we subtract using the pictures below.

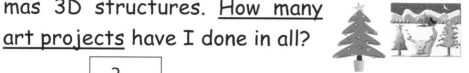

$13 - 9 = 13 - ... - ... = ...$

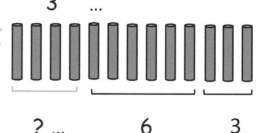

3. Find 3 rectangles in the shape and write the letters, for example, FKDB, in the box.

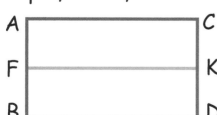

1. ... ...
2. ... ...
3. ... ...

Write the letters clockwise or counter-clockwise, but not diagonally☺.

1. I go to bed at ⟨9 pm⟩ and wake up at ⟨7 am⟩. Dad goes to bed later and wakes up at ⟨6 am⟩. When I asked him when he went to bed, he said he was sleeping ⟨2-and-a-half hours⟩ less than me. <u>When</u> does he usually go to bed? <u>Draw</u> the diagrams for the problem.

I go to bed	I wake up	I sleep	Dad wakes up	Dad sleeps	Dad goes to bed
... ___	... ___	? ...h...m	...	...h...m	? ... ___

Answer: _____.

2. I've cut the ⟨10-inch⟩ chain into ⟨2 equal⟩ pieces. <u>How long</u> is ⟨1⟩ part?

<u>How many cuts</u> did I make? ....

Chain (in)	Equal pieces	Length of 1 part (in)
...	...	? ...

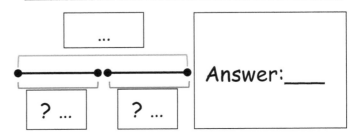

1. <u>Open</u> any book or workbook and <u>answer</u> the questions.

<u>What</u> page (p.) is before p. 9?   p. ...

<u>What</u> page (p.) is after p. 14?   p. ...

<u>Can</u> you find 2 consecutive numbers of pages if their sum is 13?

p. ... and p. ...

<u>Can</u> you find 2 consecutive numbers of pages if their sum is 16?

<u>Why</u> yes or no? _____

_____.

2. Number Sense Strategy. <u>Fill in</u> the missing numbers, <u>write</u> the number sentences, and <u>find</u> the value.

```
8 + 8           8 + 8 = ...              9 + 3           9 + ... = ...
 /\             _____               /\             _____
2   ...         8 + 2 + ... = ...        1   ...        9 + ... + ... = ...

8 + 3                                     6 + 6
 /\             _____               /\            _____
... ...                                  ... ...
```

3. <u>What bills and coins</u> may make up $6.85? <u>Write</u> 2 options.

Each quantity of bills and coins should be different.

... 5-dollar bills, ... 1-dollar bills, ... quarters, ... dimes, ... nickels, ... pennies.

... 5-dollar bills, ... 1-dollar bills, ... quarters, ... dimes, ... nickels, ... pennies.

1. Number Sense Strategy. Complete each number sentence and write in the missing numbers.

18 + 3	17 + 9	19 + 3

18 + 3 = ... + ... + ... =
... + ... = ...

17 + 9 = _____
_____

19 + 3 = _____
_____

2. There were ☐17 sunny days in September, cloudy days were ☐9 less. How many cloudy days were in September? The remaining days were rainy. How many rainy days did we have?

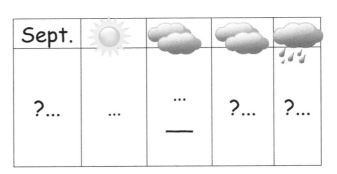

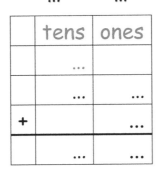

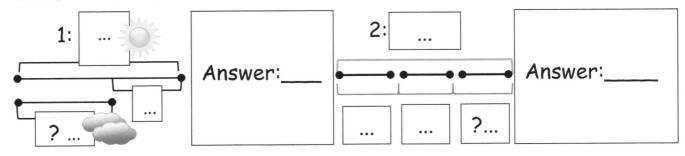

3. Who can hide in the ☐3-feet high grass? Find as many answers as you can imagine:_____
_____.

1. I have written the number ⃞11. My sister has written the number which is ⃞17 more. What number has she written? Draw the diagram.

 11 ... ... = ...

I	Sister	Sister
...	... ____	? ...

Answer:____

2. NSS. Complete each number sentence. Fill in the missing numbers.

22 - 5

	tens	ones
	...	...
	...	...
-		...
	...	...

22 - 5 = ... - ... - ... =
... - ... = ...

22 - 9

	tens	ones
	...	...
	...	...
-		...
	...	...

22 - 9 = _____
_____

23 - 6

	tens	ones
	...	...
	...	...
-		...
	...	...

23 - 6 = _____
_____

3. Compare the longest and the shortest sides of the shapes. Write the difference in the box.

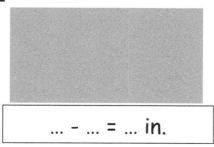
... - ... = ... in.

... - ... = ... in.

1. <u>Check</u> the scale balances. If you find a mistake, <u>write</u> the correct number in the parenthesis.

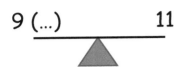  

 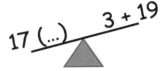

  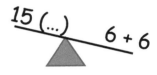

2. <u>Explain</u> how we subtract using the pictures below.

17

17 − 8 = 17 − ... − ... = ...

7 ...

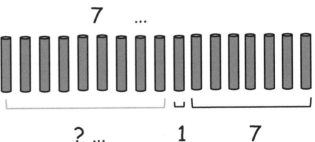

? ...   1   7

3. <u>Find</u> 5 rectangles in the shape and <u>write</u> the letters, for example, LNKC, in the box.

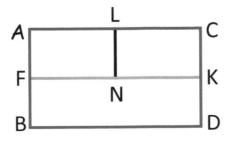

1. ... ...   2. ... ...

3. ... ...   4. ... ...

5. ... ...

1. <u>Compare</u> the longest and the shortest sides of the shapes. <u>Write</u> the difference in the box. <u>Find</u> 1 mistake and <u>cross</u> it out.

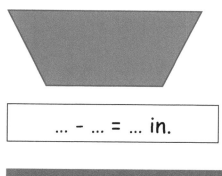

... - ... = ... in.   ... - ... = ... in.

... - ... = ... in.   ... - ... = ... in.

2. We had some rose bushes in the backyard. Mom bought $\boxed{3}$ more and now we have $\boxed{11}$ rose bushes. <u>How many bushes</u> did we have before?

? = X

X + ... = ...   X = ... - ...

X = ...

Answer:_____.

3. <u>How can</u> a giant of $\boxed{10\text{-feet}}$ high come into your room? <u>Imagine</u> as many ways as you can:_____
_____.

1. Both of my Grandpas breed bees and harvest honey. Last year Mom's Pa made ⌊32⌋ lb. of honey, and my Father's Pa made ⌊9⌋ lb. ⌊more⌋. How many lb. of honey did they make in all?

Mom's Pa	Dad's Pa	Dad's Pa	In all
...	... ___	? ...	? ...

1: ? ...   Answer:____   2: ? ...   Answer:____

Circle the correct one number sentence for the problem:

32 + 32 + 9   32 + (32 + 9)

2. Number Sense Strategy. Fill in the missing numbers, write the number sentences, and find the value. Use Round-Up-Strategy.

19 + 7   19 + 7 = ...   17 + 4   17 + ... = ...
 ∧   _____   ∧   _____
10   3   19 + 10 − 3 = ...   10   ...   17 + ... − ... = ...

                  15 + 7
                    ∧   _____
                  ...   ...

3. Draw ⌊2⌋ lines with a ruler so that the square will be divided into ⌊4 equal⌋ ⌊triangles⌋.

1. Find the value.

$\frac{2}{3} + \frac{3}{3} = \frac{...}{...}$ $\frac{1}{3} + \frac{2}{3} = \frac{...}{...}$ $\frac{2}{3} + \frac{2}{3} = \frac{...}{...}$

$\frac{1}{3} + \frac{1}{3} = \frac{...}{...}$ $\frac{1}{3} + \frac{4}{3} = \frac{...}{...}$ $\frac{3}{3} + \frac{3}{3} = \frac{...}{...}$

$1 + \frac{3}{3} = \frac{...}{...}$ $3 + \frac{1}{3} = \frac{...}{...}$ $2 + \frac{2}{3} = \frac{...}{...}$

1/3					

2. You need the scales and an apple. The color or size do not matter☺.

Take an apple and weigh it, and write down the weight. Then, cut it into 4 pieces and leave the pieces under the sun (or in the oven, but you need the help of an adult and cannot do it on your own) to dry it out. When it's dried out, take the dried pieces together and weigh them. Write down the weight.

Weight before: _____   Weight after: _____

What did you notice? _____.

How can you explain it? _____
_____.

3. Find the value. Remember: 1 hour = 60 minutes.

```
  4 h  2 0 m       5 h  4 0 m       2 h  1 5 m
+ 2 h  1 5 m     + 2 h  3 0 m     + 7 h  4 5 m
  … h  … … m       … h  … … m       … h  … … m
```

1. Number Sense Strategy. Complete or write each number sentence and fill in the missing numbers.

16 + 5

	tens	ones
	...	
	...	...
+		...
	...	...

17 + 4

	tens	ones
	...	
	...	...
+		...
	...	...

19 + 5

	tens	ones
	...	
	...	...
+		...
	...	...

16 + 5 = ... + ... + ... = ... + ... = ...

17 + 4 = _____ _____

19 + 5 = _____ _____

2. Draw the minute and hour hands red for 7:05pm. I usually go to sleep 1 hour and 45 minutes later.

Draw the minute and hour hands black for when I go to bed. _____.

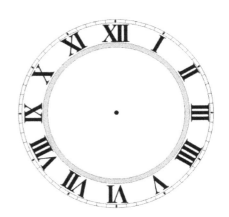

3. I had 23 bricks. How many bricks have I added if I have 32 bricks now?

Had	Have	Added
...	...	? ...

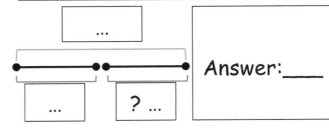

Answer:____

1. Number Sense Strategy. Complete each number sentence and write in the missing numbers.

23 - 4

3   1

	tens	ones
	...	...
	...	...
-		...
		...

23 - 4 = ... - ... - ... = ... - ... = ...

23 - 9

...   ...

	tens	ones
	...	...
	...	...
-		...
		...

23 - 9 = ... - ... - ... = ... - ... = ...

23 - 7

...   ...

	tens	ones
	...	...
	...	...
-		...
		...

23 - 7 = _____

_____

2. Take a ruler and a good mood to answer the questions.

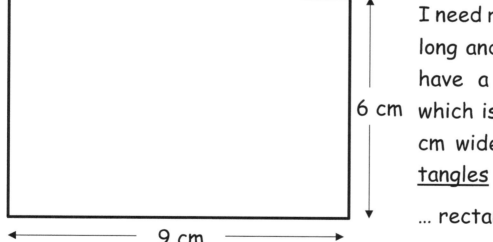

I need rectangles of 1 cm long and 3 cm wide. If I have a sheet of paper which is 9 cm long and 6 cm wide, how many rectangles can I cut out?

... rectangles.

Draw the dotted lines on the rectangle to show your cutting lines.

3. Continue the pattern:

1, 6, 11, ..., ..., ..., ...   72, 69, 66, ..., ..., ..., ...

"Draw the minute and hour hands red for 9:30am. The clock is 5 minutes behind or slow." Hm... Behind? Slow? The clock is hiding behind WHAT?! Don't you think it's scary? Definitely, creepy... Oh, I am so scared of math.

Calm down! Nothing scary!

Sometimes the clock goes faster, sometimes it goes slower. It's a kind of broken clock, that's all! If you cannot fix it, just remember: Imagine that you are walking and I am walking behind you or I am slower. What do I need to do to catch you? Go faster and add more steps.

Do you mean I need to add the minutes "behind" to the time?

As 9:30 + 0:05 = 9.35?

Perfect! When the clock is slow or behind, you just need to add the minutes to the time on the clock and you get the right time!

1. <u>Draw</u> the minute and hour hands red for 3:15 pm.

The clock is 10 minutes behind or slow.

<u>What</u> time is it now?   … : … _____.

<u>Draw</u> the minute and hour hands black.

2. <u>Write</u> the number words.

458 <u>Four hundred fifty-eight</u>_____

746 _____

583 _____

837 _____

3. I had some bricks. My sister added 8 bricks, and now I have 26 bricks. <u>How many bricks</u> did I have? <u>Draw</u> the scheme for the problem.

Had had	Added	Have	Had had
…	…	…	? …

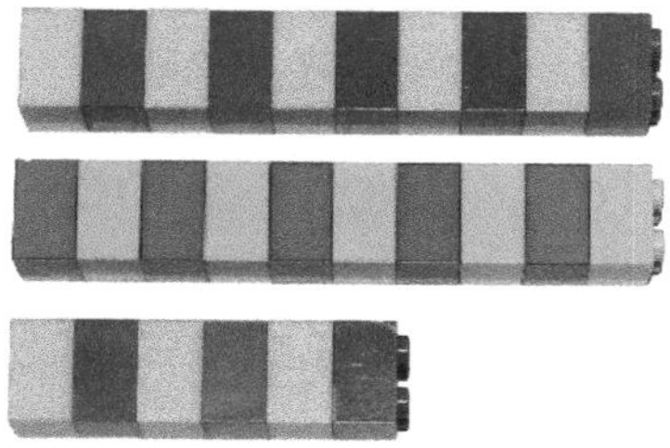

Answer:_____

"I've counted 25 apples on 2 plates. The red apples were 5 more than the green apples. How many red apples were there? How many green apples were there?"

I wonder, how can we find anything in the problem, if we do not know how many apples were of at least one kind? Red? Green? Everything is unknown.

Ha! The mistake! What's unknown? They forgot to write about red or green apples. It's simple, freaky-brainers! I quit! I'm not into problems with mistakes!

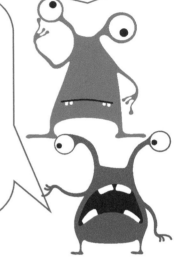

I'm not sure...Mistake?.. Strange, don't you think?.. Let's make a table: Apples in all are 25.

Red are 5 more.

Green is unknown, so, I will put "X".

Apples in all	Red	Green	Red	Green
25	5 more	X	? ...	? ...

So, we need to find red and green apples. I put "?..." for both.

Cool. I like X's. I'm writing a story about spooky-weird-cranky voids of Egypt's pyramids, scientists, pharaohs, X-ray weapons...

 "Not now! Prooooooblem! I will do the diagram. If we talk about more or less, we use comparison diagrams, right?"

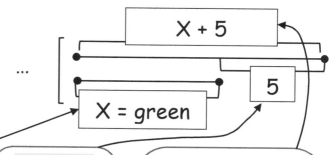

"So, the smaller amount belongs to the green apples, which are "X" for us. I put "X = green"."

"The red apples are "5 more", so, put "5" in the more/less box."

"The biggest amount shows the red apples. So, it equals the smaller amount and the difference: "X + 5"."

 "Wait, wait. Quit! I want to work, too. I've an idea! I'm full of ideas. My head is spinning how many ideas I have in my head in each moment. I think, that…"

"Disappear! Listen, dudes, I put a left bracket to connect the red and green apples, because we know that there are 25 apples in all. So, "X + 5 + X = 25"."

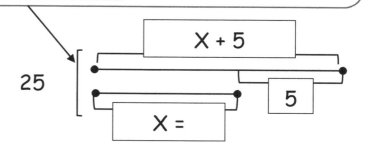

Perfect! Now, let me finish!

We can add red and green apples together:

X + X + 5 = 25

As X + X = 2X, then

2X + 5 = 25

2X = 25 − 5 ; 2X = 20

As 2X=20, 1 X is a half of 20: 10 + 10 = 20

(10 green apples)

Red: X + 5 = 10 + 5 = 15

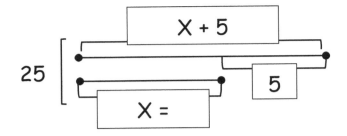

Ap-ples in all	Red	Green	Red	Green
25	5 more	X 10	? 15	? 10

X + X + 5 = 25

2X + 5 = 25   2X = 25 − 5

2X = 20

10 + 10 = 20; 10 green apples

X + 5 = 10 + 5 = 15 (red)

Answer: 10 (green), 15 (red).

1. Draw 2 lines with a ruler so that the square will be divided into 4 equal squares.

1. Last month we started a swim team. There were 42 kids, and the boys were 12 more than the girls. How many boys were on the team? How many girls were on the team?

Kids in a team	Boys	Girls	Boys	Girls
...	...	X	? ...	? ...

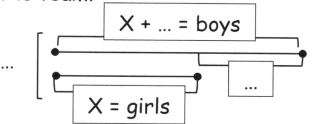

Answer:_____

2. I put 3 sticks to show the race track. The interval between sticks was 20 steps. How many steps did I take?

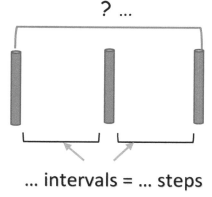

... intervals = ... steps

Answer:_____.

Why does 1 interval need 2 sticks, but 2 intervals need 3 sticks? _____
_____.

3. I decided to measure the length of my room with my steps. I counted 15 steps. My big brother counted 11 steps. My Dad counted 9 steps. How can it be? We used the steps but got different numbers: _____
_____.

What would you use to measure the room? _____.

"I need to find the value of 11 − 3 = ?.... Who wants to help?"

"I want to! I know, my way is weird, but I like it!"

11 − 3 = …

"11 is 3 + 8. So, 3 + 8 − 3 = 8. Amazing, ah?"

"Correct! We need to distribute 11 into 2 numbers, one of which is the number you subtract (3)."

15 − 8 = …

"I'll try with this one, ok?"

"15 is 8 + 7. So, 8 + 7 − 8 = 7. I did it!!!"

1. Number Sense Strategy. Write in the missing numbers and find the value.

"I'll distribute 12 into 5 and…. Help me!!!"

12 − 5 = …
|
12 is 5 + ….
So, 5 + … − … = …

17 − 8 = …
|
17 is 8 + ….
So, 8 + … − … = …

2. Fill in the missing numbers hidden behind the bricks:

… + … = 10   … + … = 15

1. Number Sense Strategy. I added two addends to ⑦ and got ㉖. Fill in the missing numbers in the number sentences to make them true.

7 + ... + ... = 26   7 + ... + ... = 26

7 + ... + ... = 26   7 + ... + ... = 26

Cross out or circle 7 bricks and play with the rest!

2. Number Sense Strategy. Fill in the missing numbers, write the number sentences, and find the value.

19 + 6   19 + 6 = ...   19 + 7   19 + ... = ...
  ∧   ─────────────   ∧   ─────────────
 1 ...   19 + 1 + ... = ...   19 + ... + ... = ...
                                       ... ...

19 + 5   ─────────────   16 + 5   ─────────────
  ∧   ∧
 ... ...

3. Draw the minute and hour hands red for ⑤:㉚ am. The clock is ⑮ minutes behind/slow.

What time is it now?   ... : ... .

Draw the minute and hour hands black.

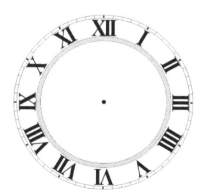

1. We bought peaches in a box. We ate 9 peaches in the evening. 16 peaches were left. How many peaches did we buy?

Peaches	Ate	Left	Buy
_____	...	...	?...

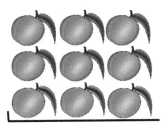 ? ... _____

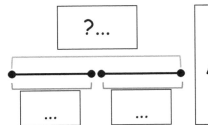

Answer:____

2. I need a gallon of water in each pot. How many more cups do I need to add to each pot to equal one gallon? Fill in each box.

1 gallon=16 cups

5 cups + ... cups

8 cups + ... cups

9 cups + ... cups

4 cups + ... cups

3. I had 45 bricks. How many bricks have I added if I have 75 bricks in all?

Had	Have	Added
...	...	?...

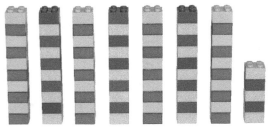

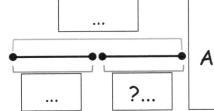

Answer:____

1. We bought tomatoes in a box. We put 7 tomatoes in the salad. 14 tomatoes were left. How many tomatoes did we buy?

Tomatoes	Salad	Left	Buy
___	...	...	?

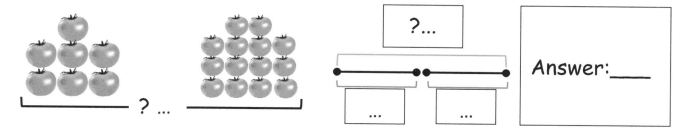

2. Number Sense Strategy. Fill in the missing numbers and find the value. Use Round-Up-Strategy.

21 – 8 = ...
|
21 is 10 - 2
So, 21 - 10 + 2 = ...

24 – 7 = ...
|
24 is ... - ....
So, ... - ... + ... = ...

3. Draw the minute and hour hands red for 7:45 am. The class starts in 25 minutes.

What time will the bell ring?   ... : ... _____.

Draw the minute and hour hands black.

4. The shape was turned 2 times to the left. Color the rectangles in their new position.

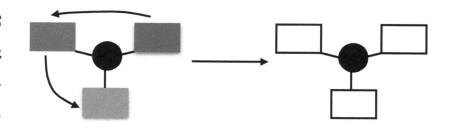

1. There were ☐25☐ episodes in the movie. I have watched ☐7☐ episodes. How many episodes were left unwatched?

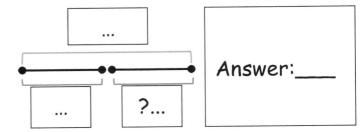

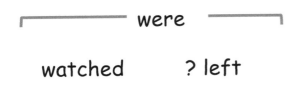

2. There were several episodes in the movie. I have watched ☐17☐ episodes and ☐8☐ more were left. How many episodes were there?

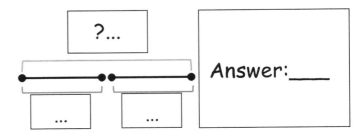

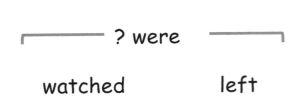

3. Number Sense Strategy. Explain how we subtract using the pictures below.

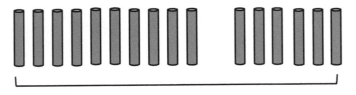

16 − 9 = 16 − ... − ... = ...

6 ...

? ...   3   6

1. Number Sense Strategy. Fill in the missing numbers and write the number sentences. Find the value.

28 + 7   28 + 7 = …   29 + 4   29 + … = …
∧   28 + 2 + … = …   ∧   29 + … + … = …
2 …   … …

38 + 4   _____   27 + 6   _____
∧   ∧
… …   … …

2.   Fill in the missing numbers.

equals 5/8 as   equals …/…

equals 3/4 as   equals …/…

3.   Open any book or workbook and answer the questions.

What page (p.) is before p. 18?   p. …

What page (p.) is after p. 25?   p. …

Can you find 2 consecutive numbers of pages if their sum is 19?

                                          p. … and p. …

Can you find 2 consecutive numbers of pages if their sum is 14?

Why yes/no? _____.

1. Number Sense Strategy. I added three addends to 5 and got 30. Fill in the missing numbers in the number sentences to make them true.

5 + ... + ... + ... = 30   5 + ... + ... + ... = 30

5 + ... + ... + ... = 30   5 + ... + ... + ... = 30

5 + ... + ... + ... = 30   5 + ... + ... + ... = 30

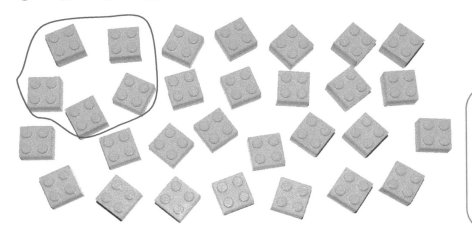

I circle 5 bricks and arrange the leftover bricks in any order.

2. There are 17 miles between my home and the swimming pool. We have already driven several miles and 9 miles are left. How many miles have we already driven?

Miles in all	Have already driven	Are left	Have already driven
...	_____	...	? ...

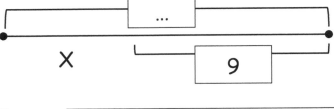

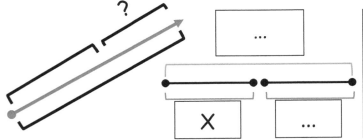

X + ... = ...

X = ...   X = ...

Answer: _____

1. Number Sense Strategy. Explain how we subtract using the pictures below.

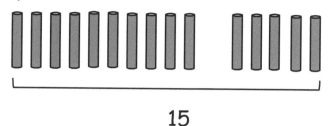

15

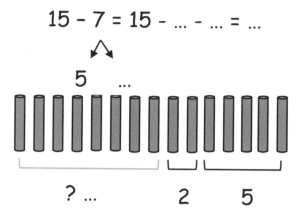

15 – 7 = 15 - ... - ... = ...

5   ...

? ...   2   5

2. Number Sense Strategy. Fill in the missing numbers and write the number sentences. Find the value. Use Round-Up-Strategy.

24 – 6 = ...
/\ ⇨   ... – ... = ...
10 ...   ... – ... + ... = ...

31 – 5 = ...
/\ ⇨   _____
... ...   _____

3. Write any [2] consecutive numbers. ..., ...

Write any [3] consecutive numbers. ..., ..., ...

Write any [4] consecutive numbers. ..., ..., ..., ...

Write any [5] consecutive numbers. ..., ..., ..., ..., ...

4. Find the value.

```
    ...                        ...
   6 h 5 0 m       8 h 1 0 m       1 h 3 2 m
 + 2 h 1 0 m     +   h 3 9 m     + 4 h 3 4 m
   _____       _____       _____
   ... h ... m     ... h ... m     ... h ... m
```

1. I've cut the 12-inch log into 2 equal pieces. How long is 1 part?

How many cuts did I make? ….

Log	Equal pieces	Length of 1 part
…	…	? …

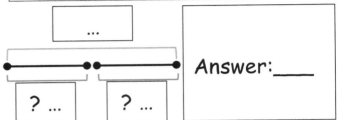

Answer:____

2. Number Sense Strategy. Complete each sentence and fill in the missing numbers. Use Round-Up-Strategy.

28 + 8
↙   ↘
10   2

	tens	ones
	…	
	…	…
+		…
	…	…

28 + 8 = … + … + …
= … - … = …

9 + 49
↙   ↘
…   …

	tens	ones
	…	
	…	…
+		…
	…	…

9 + 49 = _____
_____

4 + 39
↙   ↘
…   …

	tens	ones
	…	
	…	…
+		…
	…	…

4 + 39 = _____
_____

3. Find the value.

```
    1 $ 3 6 ¢
  + 9 $ 1 2 ¢
  ─────────────
    … … $ … … ¢
```

```
    6 $ 2 2 ¢
  + 2 $ 2 7 ¢
  ─────────────
    … $ … … ¢
```

```
        …
    7 $ 4 5 ¢
  + 1 $ 2 5 ¢
  ─────────────
    … $ … … ¢
```

1. I had bricks. I shared 9 bricks with my friend and now I have 15. How many bricks did I have before sharing?

Had	Shared	Have	Had
___	...	...	? ...

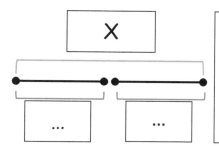

X - ... = ...

X = ...   X = ...

Answer: _____.

2. Number Sense Strategy. Explain how we subtract using the pictures below.

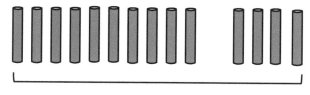

14

$14 - 8 = 14 - ... - ... = ...$

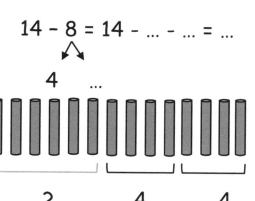

? ...   4   4

3. I want to buy a book for $9.50, and a set toys for $6. I found in my piggy bank: 2 5-dollar bills, 11 1-dollar bills, 12 quarters, 5 dimes, 4 nickels, and 100 pennies.

How much did I have in my piggy bank?

_____.

How much will be left? _____.

1. Number Sense Strategy. Fill in the missing numbers and write the number sentences. Find the value. Use Round-Up-Strategy.

26 – 9 = …
|
26 is 10 – ….

So, 26 – … + … = …

21 – 6 = …
|
21 is _____

So, _____

2. I had some games on the iPad. I added 9 games, and now I have 28 total. How many games did I have before?

Had	Added	Have	Had
___	…	…	? …

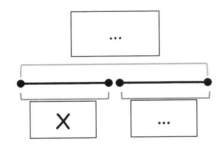

X + … = …

X = … – …   X = …

Answer: _____.

3. Answer the questions.

How many ends are on 1 log of a tree?   ….

How many ends are on 3 logs?   ….

How many ends are on 2 logs and a half?   ….

1. Number Sense Strategy. Complete each sentence and fill in the missing numbers. Use Round-Up-Strategy.

24 - 5

10   5

	tens	ones
	...	...
	...	...
-		...
	...	...

24 - 5 = 24 - 10 + ...
= ... + ... = ...

34 - 7
...   ...

	tens	ones
	...	...
	...	...
-		...
	...	...

34 - 7 = _____
_____

44 - 9
...   ...

	tens	ones
	...	...
	...	...
-		...
	...	...

44 - 9 = _____
_____

2. Explain how we subtract using the pictures below.

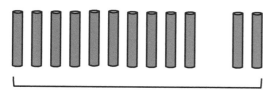

12

12 – 7 = 12 - ... - ... = ...

2   ...

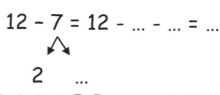

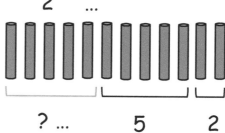

? ...   5   2

3. Find the missing of ⬚2⬚ consecutive numbers. ...,   54.

Find the missing of ⬚3⬚ consecutive numbers. 26,   ..., ...

Find the missing of ⬚4⬚ consecutive numbers. ...,   ...,   11, ...

Find the missing of ⬚5⬚ consecutive numbers. ...,   ...,   ...,   ...,   64.

1. Number Sense Strategy. I added two addends to 9 and got 29. Fill in the missing numbers in the number sentences to make them true.

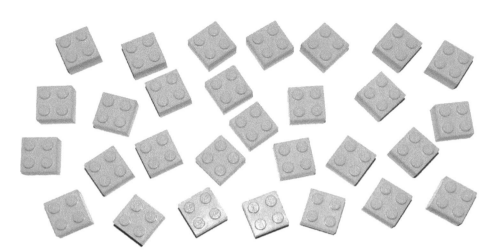

9 + ... + ... = 29

9 + ... + ... = 29

9 + ... + ... = 29

9 + ... + ... = 29

9 + ... + ... = 29

2. Find the sum or difference.

400 + 200 = 600   270 + 510 = ...   705 + 100 = ...

637 + 300 = ...   300 + 520 = ...   150 + 440 = ...

490 – 220 = ...   680 – 160 = ...   275 – 130 = ...

957 – 320 = ...   759 – 340 = ...   407 – 100 = ...

3. Answer the question.

Draw the minute and hour hands red for 11:40 am. The science class lasts 45 minutes.

When will it be over?   ... : ... _____.

Draw the minute and hour hands black.

4. Continue the pattern: 4, 8, 12, ..., ..., ..., ..., ..., ...

1. I've counted 85 roses on 2 bushes. The red roses were 25 more than the pink roses. How many red roses were there? How many pink roses were there?

Roses in all	Red	Pink	Red	Pink
...	...	X	? ...	? ...

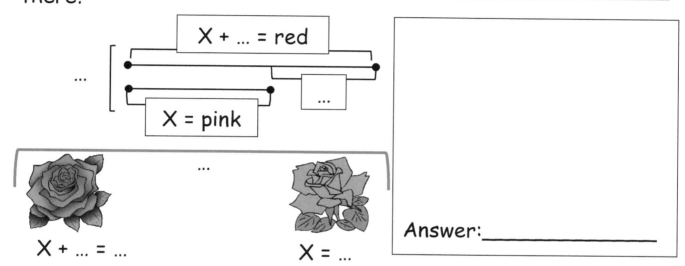

X + ... = red
X = pink
X + ... = ...
X = ...

Answer: _____

2. Find 8 triangles in the shape and write the letters, for example, BKD, in the box. The order of the letters does not matter.

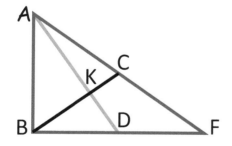

1. ... ...   2. ... ...
3. ... ...   4. ... ...
5. ... ...   6. ... ...
7. ... ...   8. ... ...

3. Divide the rectangle into equal 4 parts where each part has the same number of squares and 1 circle.

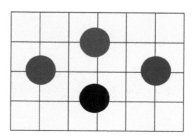

1. I am as old as my sister and my sister is older than my brother. My cousin is older than my sister. Draw the diagram to show the age of kids in our family from the oldest to the youngest. (For example, I am older than my brother show as I > Brother).

   _____   _____   _____   _____

2. Answer the questions.

   Sketch a circle in the box and divide it into 5 equal parts. Compare the fractions, using <, >, or =:

   $\frac{1}{5}$ ... $\frac{4}{5}$? $\frac{3}{5}$ ... $\frac{2}{5}$?

3. Number Sense Strategy. Complete each sentence and fill in the missing numbers. Use Round-Up-Strategy.

   27 + 6   6 + 49   5 + 77

	tens	ones
	...	
	...	...
+		...
	...	...

   27 + 6 = ... + ... + ...
   = ... - ... = ...

	tens	ones
	...	
		...
+	...	...
	...	...

   6 + 49 = _____

	tens	ones
	...	
		...
+	...	...
	...	...

   5 + 77 = _____

1. I've cut the 12-inch chain into 3 equal pieces. How long is 1 part?

How many cuts did I make? ….

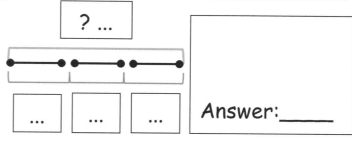

Chain	Equal pieces	Length of 1 part
…	…	? …

Answer:_____

2. Write in the missing numbers.

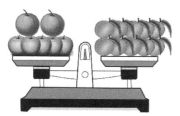

5 apples + 2 oranges = 10 peaches

2 oranges = 5 peaches

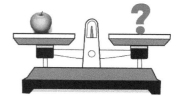

1 apple = ? … peach(es)

3. Color the shapes so that the square is on top of the circle, and the triangle is on top of the square.

The red circle is on top of the yellow triangle since you can see the whole circle. The yellow triangle is on top of the green square because you can see more of the triangle than of the square.

1. Number Sense Strategy. I added two addends to ⌑8⌑ and got ⌑30⌑. Fill in the missing numbers in the number sentences to make them true.

8 + ... + ... = 30

8 + ... + ... = 30

8 + ... + ... = 30

8 + ... + ... = 30

8 + ... + ... = 30

8 + ... + ... = 30

8 + ... + ... = 30

2. NSS. Complete each sentence. Fill in the missing numbers.

38 + 7   54 + 8   88 + 9

10   3   ...   .

tens	ones
...	
...	...
+	...
...	...

tens	ones
...	
...	...
+	...
...	...

tens	ones
...	
...	...
+	...
...	...

38 + 7 = ... + ... − ... =

... − ... = ...

54 + 8 = _____

88 + 9 = _____

3. Cross out what does not belong.

1. There were 25 pages in the book. I have read 8 pages. <u>How many pages</u> were left?

Were	Read	Left
...	...	?...

In all: ... pages

Read: ... pages

Left: ?... pages

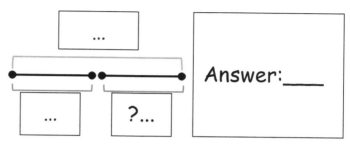

Answer:____

2. There were some pages in the book. I have read 9 pages and 16 pages were left. <u>How many pages</u> were there?

Were	Read	Left	Were
____	...	...	?...

In all: ?... pages

Read: ... pages

Left: ... pages

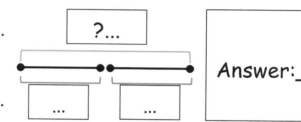

Answer:____

3. Number Sense Strategy. <u>Fill in</u> the missing numbers and write the number sentences. <u>Find</u> the value. The first one is done for you.

24 – 7 = ...
/\
10   14
⇒ 10 – 7 = ...
   ... + 14 = ...

33 – 8 = ...
/\
10   23
⇒ 10 – 8 = ...
   ... + 23 = ...

52 – 5 = ...
/\
...   ...
⇒ ... – ... = ...
   ... + ... = ...

81 – 4 = ...
/\
...   ...
⇒ _____
  _____

1. Number Sense Strategy. Fill in the missing numbers and write the number sentences. Find the value.

47 + 6   47 + 6 = …   78 + 6   78 + … = …
 /\   _____   /\   _____
3   3   47 + … + … = …   78 + … + … = …
                                    …   …

29 + 8   _____   59 + 5   _____
 /\   /\
…   …   …   …

2. You need a scale, 3 tablespoons of salt, a cup of water and a large shallow saucer or a plate, and a good mood ☺.

Take 3 tablespoons of salt, weigh it, and write down the weight.

Then, dissolve it into a half cup of water, pour it onto a large shallow saucer or a plate, leave it under the sun to evaporate (or ask for an adult to help you boil it in a pot until it's evaporated). When all the water is evaporated, pick up the leftover salt and weigh it. Write down the weight.

Weight before: _____   Weight after: _____

What did you notice? _____.

How can you explain it? _____
_____.

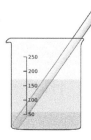

1. <u>Explain</u> how we subtract using the pictures below.

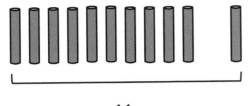

11 – 6 = 11 - ... - ... = ...

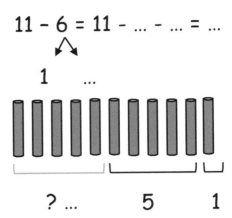

2. I had ⎡7⎤ red bricks. Then, I added ⎡14⎤ blue and ⎡9⎤ green bricks. <u>How much more</u> did I add to what I had?

Had	Ad-dends	Added	Added > had
...	..., ...	?...	?...

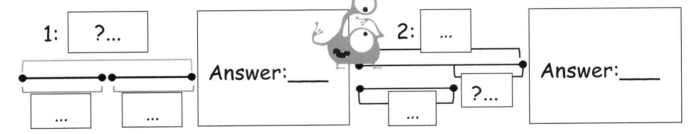

3. <u>Find</u> the missing of ⎡2⎤ odd consecutive numbers. ..., 45.

<u>Find</u> the missing of ⎡3⎤ odd consecutive numbers. 13, ..., ...

<u>Find</u> the missing of ⎡4⎤ odd consecutive numbers. ..., ..., 21, ...

<u>Find</u> the missing of ⎡5⎤ odd consecutive numbers. ..., ..., ..., ..., 39.

4. The shape was turned ⎡2⎤ times to the right. <u>Color</u> the rectangles in their new position.

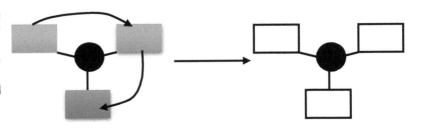

1. Write the number sentence for the problem. Find the value.

83621: Find the sum of the thousands and tens _____, the difference of the hundreds and ones _____, the sum of the ten thousands, hundreds and tens _____, the difference of the ten thousands, thousands and ones _____.

2. I've counted 14 sailors on 2 boats. The white boat had 2 more sailors than the blue boat. How many sailors were on the white boat? How many sailors were on the blue boat?

Sailors in all	White	Blue	White	Blue
...	...	X	?	?
	____		...	...

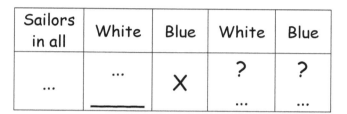

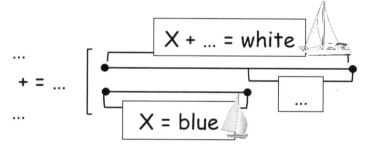

Answer:_____

3. Follow the directions and color the circles.

Start with the brown circle: go 3 circles to the right and 3 circles down. Color the circle green.

Then, go 2 circles to the left, 2 circles up, and 3 circles to the right. Color the circle yellow.

1. Number Sense Strategy. Complete each sentence and fill in the missing numbers.

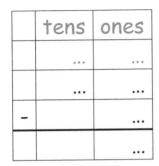

34 - 8

4   4

34 - 8 = … - … - … =
… - … = …

45 - 7

…   …

45 - 7 = _____
_____

25 - 9

…   …

25 - 9 = _____
_____

2. Number Sense Strategy. I added two addends to 6 and got 31. Fill in the missing numbers to make the number sentences true.

6 + … + … = 31   6 + … + … = 31

6 + … + … = 31   6 + … + … = 31

6 + … + … = 31

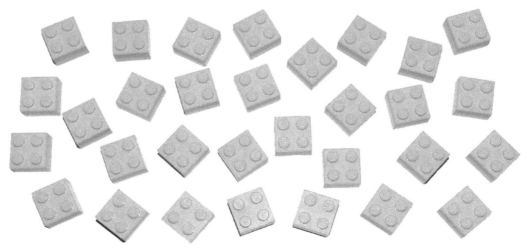

Do you know, TANGRAM which means seven boards of skill, was supposed to be invented in China and brought to Europe in the 19th century? Cut out the colored shapes and be ready to play!

You can build different shapes with the pieces of this amazing puzzle. Let's try with the colored shapes.

1. I had 9 green bricks, then, I added 7 yellow and 5 red bricks. How much more did I add to what I had? Draw the diagrams.

Had	Ad-dends	Added	Added > had
...	..., ...	? ...	?

1:

Answer:____

2:

Answer:____

2. Number Sense Strategy. Fill in the missing numbers and find the value. The first one is done for you.

12 − 5 = ...

12 is 5 + 7.

So, 5 + ... − ... = ...

17 − 8 = ...

17 is ... + ...

So, _____

3. My 2 sisters (Anna and Mary) and I (Lisa) were playing "Scary Woods". We chose 3 creatures: a dragon, a fairy, and a princess. Mary and I were not princesses. Mary is also a romantic, she even sleeps with a magic wand. Find out what creatures we chose. Write Y for Yes and N for No.

	Dr.	F.	Pr.
A.	...	...	...
M.	...	...	...
L.	...	...	...

4. Find X. Remember: Part1 = Whole − Part2.

X + 200 = 700

X =

X + 400 = 900

X =

X + 100 = 700

X =

1. Grandma gave ⬚18⬚ apples. We put ⬚7⬚ apples into apple pie. How many apples were left? Draw the diagram.

Gave	Put	Left
...	...	?...

Answer:___

2. Grandma gave us apples. We put ⬚7⬚ apples into an apple pie. ⬚11⬚ apples were left for jam. How many apples did she give? Fill in the table.

Apple pie

Jam

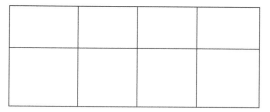

Answer:___

3. Number Sense Strategy. Fill in the missing numbers and write the number sentences. Find the value.

$24 - 8 = ...$
  /\
10   14
⇒
$10 - 8 = 2$
$14 + 2 = ...$

$23 - 5 = ...$
  /\
10   ...
⇒
$... - ... = ...$
$... + ... = ...$

$33 - 6 = ...$
  /\
...   ...
⇒
$... - ... = ...$
$... + ... = ...$

$26 - 9 = ...$
  /\
...   ...
⇒
$... - ... = ...$
$... + ... = ...$

1. Number Sense Strategy. Write in the missing numbers and the number sentences. Find the value.

13 – 6 = ...
|
13 is 6 + 7

So, 6 + 7 – 6 = ...

15 – 6 = ...
|
15 is _____

So, _____

17 – 8 = ...
|
17 is ... + ...

So, ... + ... – ... = ...

18 – 9 = ...
|
18 is _____

So, _____

2. Fill in the missing numbers and write in the fraction.

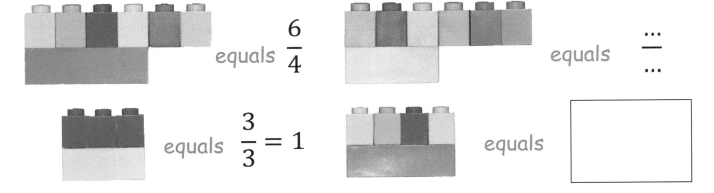

equals $\frac{6}{4}$

equals $\frac{...}{...}$

equals $\frac{3}{3} = 1$

equals

3. Fill in the missing numbers and complete the number sentences to find the value. Use Round-Up-Strategy.

17 – 9 = ...
|
17 is 10 - 1

So, 12 - ... + ... = ...

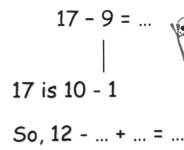

You'll see I will send a truck into space after doing all this math!

12 – 7 = ...
|
12 is _____

So, _____

"Do you remember, we always subtract ones from ones and tens from tens. You see that "20" has 0 ones."

20 - 9
↙ ↘
10 10

tens	ones
...	...
2	0
-	9
...	...

"Hmmm... But it has 2 tens."

"Aha, you can represent 20 as 10 + 10."

"Hmm.. But you cannot subtract ones digit 9 out of ones digit 0: 0 – 9 = ?!? GGRRR..."

"Wait...Then, 10 + 10 – 9 = 10 + 1 = 11. I solved it!!!"

tens	ones
...	...
2	0
-	9
...	...

"Let me clear it out. If I calculate 20 - 9 in a column, I need one more row above my problem."

"As you said, I cannot subtract 9 out of 0. But I can borrow 10 out of 2 tens. What is 20? 20 = 10 + 10. When I borrow 1 ten, I get 10 ones instead of 0 ones, and I write 10 above 0 ones.

I write 1 above 2 tens.

I cross out 2 and 0 to avoid confusion."

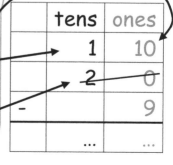

	tens	ones
	~~2~~ 1	~~0~~ 10
−		9
	1	1

Now, can I subtract 9 out of 10? Easy! 10 − 9 = 1, I write 1 in the ones.

I am left with 1 more ten and I write 1 in the tens column. Done: 11.

I hate to think a lot and count with columns. I like to do math fast or I'm getting mad, really mad!

Sure, sure, sure. Then, especially for you — the best way to solve subtraction and addition problems is to think of tens!

First, 20 = 10 + 10

Third, we add remaining 10 and 1 to get 11 (1 tens 1 ones): 10 + 1 = 11.

Second, we subtract 9 from 10 and get 1:

10 − 9 = 1.

20 − 9 = 10 + 10 − 9 =

10 + 1 = 11

1. <u>Continue</u> the pattern:

99, 88, 77, ____, ____, ____, ____, ____, ____.

1. I need 30 bricks of orange and blue colors each. If I have 20 orange bricks and 19 blue bricks, how many more bricks of each color do I need?

Need orange and blue each	Have orange	Have blue.	Need orange	Need blue
...	...	...	?...	?...

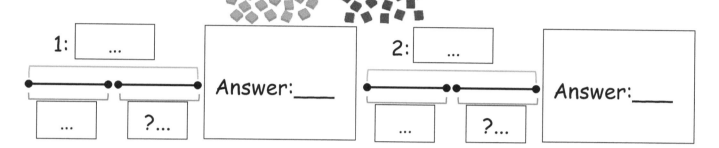

1: [...]   2: [...]

[...] [?...]   Answer:___

[...] [?...]   Answer:___

2. <u>Find</u> the missing addend. <u>Remember</u>: Whole = Part1 + Part2.

$x - 3 = 47$ $x - 2 = 78$ $x - 8 = 52$

$x =$ $x =$ $x =$

$? - 4 = 26$ $? - 6 = 44$ $? - 9 = 81$

$? =$ $? =$ $? =$

🐭 $- 5 = 65$ 🐵 $- 7 = 23$ 🐼 $- 1 = 39$

🐭 $=$ 🐵 $=$ 🐼 $=$

3. <u>Compare</u> the fractions, using >, <, or =.

$\frac{5}{6}$ ... $\frac{2}{3}$ $\frac{4}{6}$ ... $\frac{2}{3}$ $\frac{3}{3}$ ... $\frac{6}{6}$

	1/3		

1/6					

1. Number Sense Strategy. Complete each number sentence, fill in the missing numbers. Use your favorite strategy.

	tens	ones
	...	...
	...	...
-		...
	...	...

46 - 7

...   ...

46 - 7 = _____

_____

	tens	ones
	...	...
	...	...
-		...
	...	...

25 - 8

...   ...

25 - 8 = _____

_____

	tens	ones
	...	...
	...	...
-		...
	...	...

35 - 6

...   ...

35 - 6 = _____

_____

2. Find 4 triangles in the shape and write the letters, for example, FCD, in the box. The order of the letters does not matter.

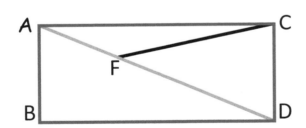

1. ... ...
2. ... ...
3. ... ...
4. ... ...

3. Find X. Remember: Whole = Part1 + Part2.

X - 200 = 300   X - 400 = 600   X - 100 = 700

X =   X =   X =

4. Cross out what does not belong.

1. <u>Find</u> the value.

$1 - \frac{2}{3} = \frac{...}{...}$ $1 - \frac{1}{3} = \frac{...}{...}$ $1 - \frac{3}{3} = \frac{...}{...} = ...$

$\frac{2}{3} - \frac{1}{3} = \frac{...}{...}$ $\frac{4}{3} - \frac{1}{3} = \frac{...}{...} = ...$ $\frac{3}{3} - \frac{2}{3} = \frac{...}{...}$

$\frac{3}{3} + \frac{1}{3} - \frac{2}{3} = \frac{...}{...}$ $\frac{4}{3} - \frac{2}{3} - \frac{1}{3} = \frac{...}{...}$ $\frac{2}{3} + \frac{3}{3} - \frac{4}{3} = \frac{...}{...}$

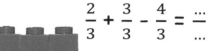

2. <u>Take</u> a ruler, <u>measure</u> the rectangle, and <u>answer</u> the questions.

<u>Divide</u> a rectangle into 6 equal parts.

<u>How many</u> is 1 part of the rectangle? $\frac{...}{...}$.

<u>What kinds of shapes</u> did you get?
_____.

3. My Grandma has chickens and rabbits. I've counted 11 heads and 32 legs.

<u>How many chickens</u> does she have? _____.

<u>How many rabbits</u> does she have?
_____
_____.

"Draw the minute and hour hands red for 9:30am. The clock is 5 minutes ahead or fast." Hm… Ahead? The clock is WHAT?! Mad, I am soooo mad!!! Why can't the clock work right? What's the problem with the clock? Let's stop doing math and take the clock to the repair shop, right?

No, no, no! Nothing to be mad about. Imagine: you are walking and I am walking ahead of you or I am faster.

I hate it when you are walking faster, I need to walk more and make more steps to catch you or ask you to slow down or freeze. All options drive me mad.

Correct! When <u>the clock is fast or ahead</u>, you just need to <u>subtract the minutes out of the time on the clock</u> and you get the right time: 9:30 – 0:05 = 8:25

1. <u>Cross out</u> the picture which does not belong.

   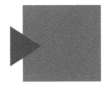

1. <u>Circle</u> the bricks which you need to fill up the rectangle and <u>cross out</u> the bricks which are not needed.

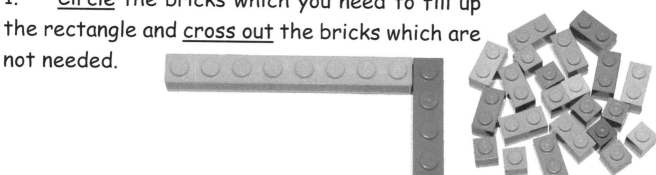

2. NSS. <u>Complete</u> the number sentences and <u>fill in</u> the missing numbers.

20 - 1

	tens	ones
	1	10
	2	0
-		1
	1	9

20 − 1 = 10 + 10 − 1 = … + … = …

20 - 3

… …

	tens	ones
	…	…
	…	…
-		…
	…	…

20 − 3 = … + … − … = … + … = …

20 - 5

… …

	tens	ones
	…	…
	…	…
-		…
	…	…

20 − 5 = … + … − … = … + … = …

3. <u>Compare</u> fractions using the bars below, <u>insert</u> >, <, or =.

$\frac{1}{2}$ … $\frac{4}{6}$ $\frac{1}{6}$ … $\frac{1}{3}$ $\frac{2}{3}$ … $\frac{4}{6}$

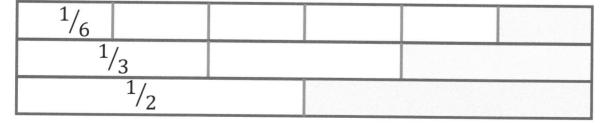

1. Find the missing of 2 even consecutive numbers. ...., 28.

Find the missing of 3 even consecutive numbers. 16, ..., ...

Find the missing of 4 even consecutive numbers. ...., ..., 32, ...

Find the missing of 5 even consecutive numbers. ...., ..., ..., ..., 50.

2. I had $76 from my birthday. I've spent $57. How much more have I spent than is left over?

Had	Spent	Left over	Spent > Left
...	...	? ...	? ...

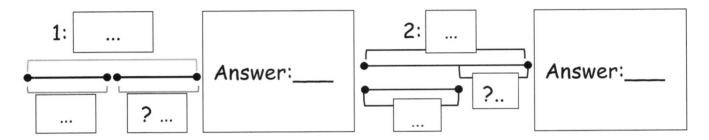

3. Fill in the missing numbers so that each side of the triangle must add up to the number in the center.

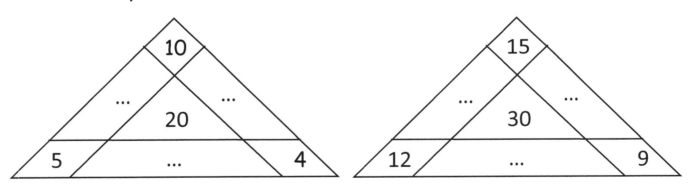

4. Find X. Remember: Part1 = Whole − Part2.

$X + 125 = 678$ $X + 375 = 595$ $X + 425 = 575$

$X =$ $X =$ $X =$

Now you can try to make the black shapes with the TANGRAM puzzle pieces.

1. It was 47°F in the morning and 55°F in the afternoon. How much warmer was it in the afternoon?

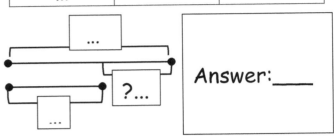

Morning	Afternoon	How much warmer?...
...	...	...

Answer:___

2. It was 47°F in the morning, it got 5° more in the afternoon. How warm was it in the afternoon?

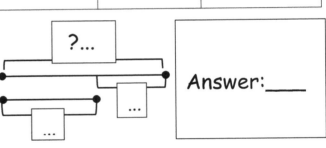

Morning	Afternoon	Afternoon
...	... ___	...

Answer:___

3. School starts at 7:45am. Draw the minute and hour hands red. Mom dropped me off at 7:15am. Draw the minute and hour hands black.

How long did I wait for the first class?

_____.

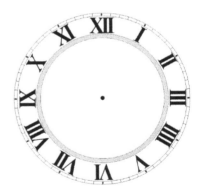

4. Find X. Remember: Whole = Part1 + Part2.

X - 125 = 673   X - 314 = 254   X - 423 = 521

X =   X =   X =

1. There are 30in. between a guitar and a chair in my little sister's castle. How many inches will be between them if I move the chair 15 inches to the right and the guitar 17 in. to the left? Draw the arrows to show my moves.

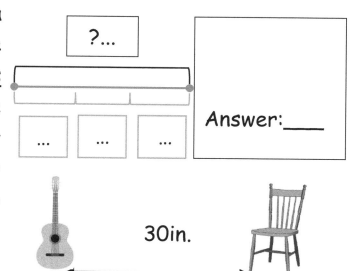

2. Color the rectangle red on the top of each blue shape.

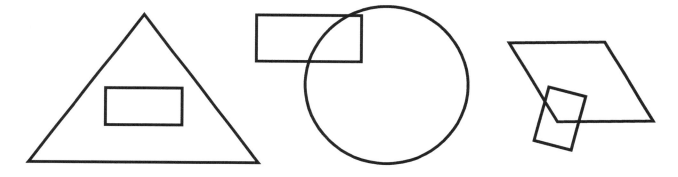

3. Write the number words.

271 _____

_____

946 _____

_____

555 _____

_____

1. I've drawn a 10cm long and 8cm high rectangle. You may look at what Briner says below ☺.

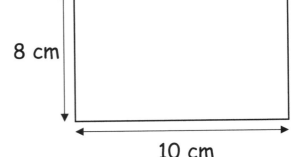

Find the Perimeter:

P = _____

Find 3 different measurements of rectangle's length and height if the Perimeter is 36 sq.cm., the length is more than the height:

Length = ... cm   Length = ... cm   Length = ... cm

Height = ... cm   Height = ... cm   Height = ... cm

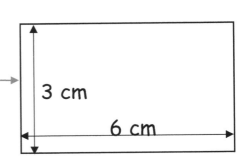

P = 3 + 3 + 6 + 6 = 18.

I know a trick! Imagine, I have a rectangle. Its Perimeter is the length around the shape. So, you need to add all the sides of a shape: 3+3+6+6=18.

Two pair of equal sides: 3cm (the height) and 6cm (the length). The sum of the length and height is: 3+6=9.

So, the sum of 2 sides in any rectangle with the perimeter of 18 should be 9. I will circle two sides.

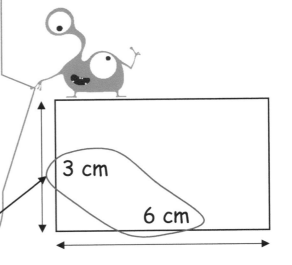

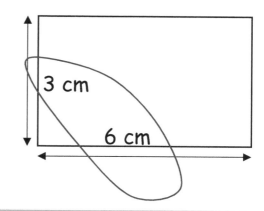

"The trick is: the sum of the length and height is always the same, but the addends are different. So, you need to remember word families for the sum."

"Word families for 9, right? 8 and 1; 7 and 2; 6 and 3; 5 and 4."

"I knew you would make a mistake! You did not study well last year. You forgot about 9 and 0. Smarty, you should always ask me. My memory is brilliant."

"Ha-ha. Where did you see a rectangle that is 0-cm long???"

"Sorry, I've a flu, you know…"

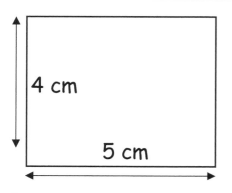

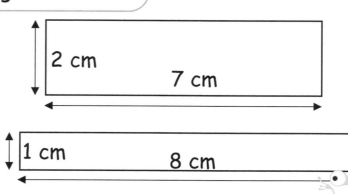

"Now, go back to p. 150 and finish your problem!"

1. Compare fractions using the bars below, fill in >, <, or =.

$\frac{1}{3}$ ... $\frac{2}{6}$ $\frac{4}{6}$ ... $\frac{3}{3}$ $\frac{1}{3}$ ... $\frac{1}{2}$

$\frac{5}{6}$ ... $\frac{3}{6}$ $\frac{6}{6}$ ... 1 $\frac{3}{6}$ ... $\frac{1}{2}$

2. I had ⬜18 problems from the Addition workbook and ⬜19 problems from the Subtraction workbook. I have solved ⬜29 of the problems. How many more should I solve?

Addition	Subtraction	In all	Solved	Need more
...	...	?...	?...	?...

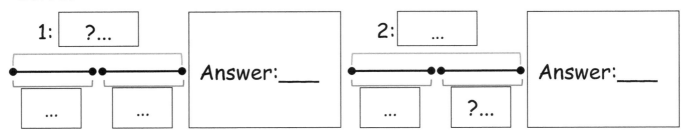

Write one number sentence for the problem:

_____.

3. Fill in the missing "+" or "-" to make the number sentences true.

a) 3 ... 3 ... 3 ... 3 = 6
b) 3 ... 3 ... 3 ... 3 ... 3 = 3
c) 3 ... 3 ... 3 ... 3 = 12
d) 3 ... 3 ... 3 ... 3 ... 3 = 15

1.

Now you can try to make the black shapes.

2. Continue the pattern:

98, 94, 88, ____, ____, ____, ____, ____, ____.

1. <u>Follow</u> the directions and <u>color</u> the circles.

<u>Start</u> with the blue circle: <u>go</u> |4| circles to the right, |2| circles down, and |3| circles to the left. <u>Color</u> the circle red.

Then, <u>go</u> |1| circle up, |2| circles to the right, |2| circles down, |3| circles to the left. <u>Color</u> the circle green.

2. I want to buy a chocolate ice cream for |$3.50|, and a toy for |$7.99|. I found in my piggy bank: |1| 5-dollar bills, |4| 1-dollar bills, |20| quarters, |8| dimes, |20| nickels, and |19| pennies.

<u>How much</u> did I have in my piggy bank?
_____.

<u>How much</u> will be left?_____
_____.

3. I have |95| bricks: |22| are red and |31| are green. <u>How many yellow</u> <u>bricks</u> do I have? <u>Draw</u> the scheme.

Bricks in all	Red	Green	Yellow
...	...	...	? ...

Answer:____

1. <u>Fill in</u> the missing numbers and <u>find</u> the value.

12 − 4 = …
|
12 is 4 + ….

So, … + … − … = …

13 − 7 = …
|
13 is … + ….

So, … + … − … = …

2. <u>Check</u> the scale balances. If you find a mistake, <u>write</u> the correct number in the parenthesis.

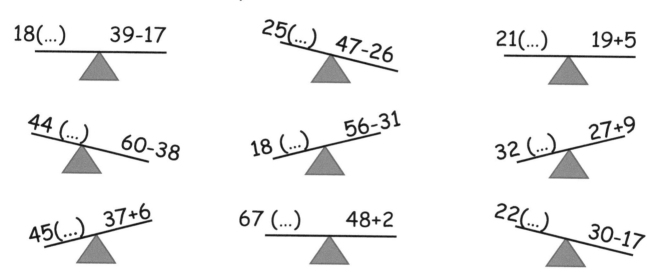

3. I subtracted ⑤ and ⑦ bricks out of ㉝ bricks. <u>How many more or less bricks</u> are left over than were subtracted?

Had	Subtracted	Subtracted	More/Less Left over
…	…, …	?...	?...

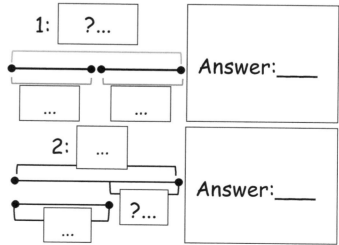

1. <u>Guess</u> the width and the length of your backyard (or school's playground, or your room). <u>Guess</u> the length as well. <u>Sketch</u> the shape in the box and <u>write down</u> your measuring guesses.

<u>Measure</u> the width and the length with a measuring tape, <u>sketch</u> the shape in the box, and <u>write down</u> the actual measurements.

Guess:	Measurements:

<u>What</u> is the difference in the widths? … - … = …_____.

<u>What</u> is the difference in the lengths? … - … = …_____.

<u>Use</u> a timer to <u>find out</u> <u>how fast</u> you can walk from one end of the backyard (or a playground, or a room) diagonally to the opposite end?   … : ….

<u>Use</u> a timer to <u>find out</u> <u>how fast</u> you can run from one end of the backyard (or a playground, or a room) diagonally to the opposite end?   … : ….

1. Use the clock to solve the problem.

I must be at school at 7:30 am ☺. I have a 30-minute lunch break at 10.30 am and a 30-minute recess at 12 am ☺. I leave school at 2:30 pm. How much time do I study at school?

Come to school	Lunch break	Recess	Leave school	Time to study in school
…:…	… …:…	… …:…	…:…	? …:…

Answer:_____.

2. My little sister likes to play tea party with her dolls. She puts them around the table. Her Ariel is always sitting the fifth from the Rapunzel doll, clockwise or counter-clockwise. How many dolls are sitting at the table? Draw the dolls.

_____.

1. I have built the tower with 17 bricks. My sister's tower has 9 less bricks. How many bricks does she have?

I	Sister	Sister
...	...	?
	_____	...

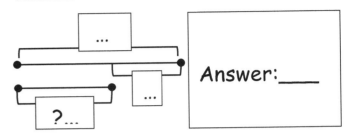

Answer:___

2. Number Sense Strategy. Fill in the missing numbers and write the number sentences. Find the value.

49 + 9   49 + 9 = ...   37 + 7   _____
 /\   _____   /\
1   ...   49 + 1 + ... = ...

3. I added 5, 8, and 3 to 4. How much more is the sum now than the number 4?

Had	Addends	Sum	Sum > had
...	..., ..., ...	?...	?...

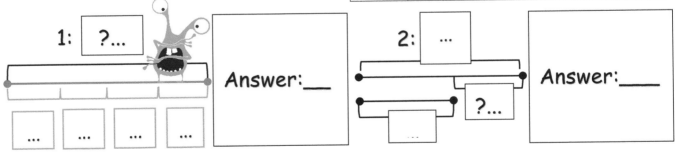

1: ?...   Answer:___   2: ...   Answer:___

... ...   ?...

4. Find X. Remember: Part1 = Whole – Part2.

X + 300 = 760   X + 200 = 580   X + 400 = 850

X =   X =   X =

1. <u>Number Sense Strategy.</u> <u>Complete</u> each number sentence and <u>fill in</u> the missing numbers to find the value. <u>Use</u> Round-Up-Strategy.

30 - 7   40 - 2   50 - 4

10   3   10   ...   .

	tens	ones
	...	...
	...	...
-		...
	...	...

	tens	ones
	...	...
	...	...
-		...
	...	...

	tens	ones
	...	...
	...	...
-		...
	...	...

30 – 7 = 30 - 10 + 3 = ... + ... = ...

40 – 2 = _____
_____

50 – 4 = _____
_____

2. <u>Color</u> the triangle yellow so that it's underneath of a green shape.

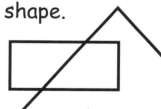
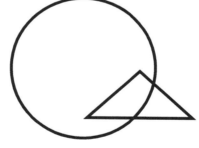
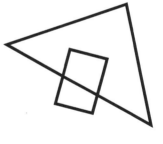

3. I have $ 55 . My brother has $ 7 more. <u>How many dollars</u> does he have?

I've $...

Brother has ? $...

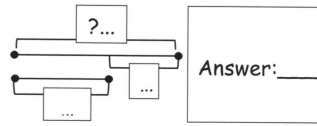

I have	Brother	Brother
...	... _____	? ...

?...

...

Answer:____

1. Number Sense Strategy. Fill in the missing numbers and write in the number sentences. Find the value.

28 – 9 = ...   10 – 9 = ...
  /\   ⇨
 10 18   ... + 18 = ...

25 – 7 = ... – ... = ...
  /\   ⇨
 10 ...   ... + ... = ...

37 – 9 = ... – ... = ...
  /\   ⇨
 ... ... + ... = ...

32 – 8 = ...   _____
  /\   ⇨
 ... ...   _____

2. If my sister and I add our candies, we will have 24 candies. If we subtract them, the difference will be 4 candies. She has more candies than I have. No wonder, she's older☹. How many candies do I have?

Sum	Difference	Sister	I
			?
...	...	___	...

Answer: _____

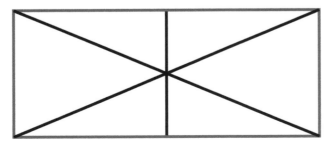

3. Find the total number of quadrilaterals and triangles of any size in these shapes.

Quadrilaterals _____

Triangles   _____

1. <u>Find out</u> <u>what number</u> is hiding behind X, ?, or an animal's face. <u>Remember</u>: X = Part1 + Part2. Add ones. Then, add tens.

X - 13 = 47	X - 12 = 78	X - 38 = 52
X = 13 + 47 = 60	X = _____	X = _____
? - 34 = 26	? - 36 = 44	? - 19 = 81
? = _____	? = _____	? = _____
🐭 - 45 = 35	🐵 - 67 = 23	🐼 - 71 = 19
🐭 = _____	🐵 = _____	🐼 = _____

2. My brother is [14] years old, I am [8] years old. <u>How old</u> will my brother be when I am [as old as] he is now? (Brother is weird, ah?).

My brother now	I am now	Difference in our ages	My brother in ... years
...	...	?...	?...

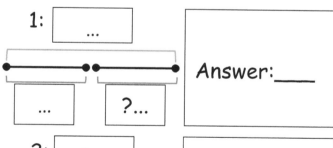

8y.o. ⟶ 14y.o ⟶ ?... y.o.

<u>Write</u> one number sentence for the problem:

_____

3. <u>Find</u> X.

X - 300 = 560	X - 200 = 580	X - 400 = 350
X =	X =	X =

1. Number Sense Strategy. I added three addends to 6 and got 28. Fill in the missing numbers in the number sentences to make them true.

6 + ... + ... + ... = 28   6 + ... + ... + ... = 28

6 + ... + ... + ... = 28   6 + ... + ... + ... = 28

6 + ... + ... + ... = 28   6 + ... + ... + ... = 28

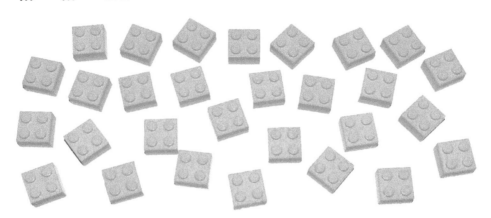

2. Answer the questions.

Mom cooked 4 steaks for dinner. There were 2 fathers and 2 sons at the table. Each of them ate 1 steak and there was 1 steak left on the plate. Is it possible? Who were they? Draw them sitting at the table.

Where did 1 son or Dad disappear?..

Who are they? They are: _____

_____.

Alphabet Table

1	A	10	J	19	S
2	B	11	K	20	T
3	C	12	L	21	U
4	D	13	M	22	V
5	E	14	N	23	W
6	F	15	O	24	X
7	G	16	P	25	Y
8	H	17	Q	26	Z
9	I	18	R		A is the 1st, Z is the 26th

ANSWERS and SOLUTIONS

1. <u>Answer</u> the questions and <u>fill in</u> the missing letters.
Below is a 7-inch long rectangle divided into 7 equal parts.

Hints: 1) <u>draw</u> a 7-inch long rectangle; 2) <u>measure</u> 7 equal parts of 1-inch long; 3) <u>draw</u> the 7 vertical black-dotted lines. You will get 7 equal parts.

<u>How many</u> is 1 part out of 7 equal parts? $1 \div 7 = \frac{1}{7}$ (one-seventh).

<u>How many</u> are 2 parts out of 7? $2 \div 7 = \frac{2}{7}$ (<u>two-sevenths</u>).
<u>How many</u> are 3 parts out of 7? $3 \div 7 = \frac{3}{7}$ (<u>three-sevenths</u>).
<u>How many</u> are 4 parts out of 7? $4 \div 7 = \frac{4}{7}$ (<u>four-sevenths</u>).
<u>How many</u> are 5 parts out of 7? $5 \div 7 = \frac{5}{7}$ (<u>five-sevenths</u>).
<u>How many</u> are 6 parts out of 7? $6 \div 7 = \frac{6}{7}$ (<u>six-sevenths</u>).
<u>How many</u> are 7 parts out of 7? $7 \div 7 = \frac{7}{7} = 1$.

The whole always equals 1 or $\frac{2}{2}$, or $\frac{3}{3}$, or $\frac{4}{4}$, or $\frac{6}{6}$, or $\frac{7}{7}$.

---

1. <u>Number Sense Strategy</u>. <u>Fill in</u> the missing numbers to complete each number sentence. <u>Use</u> the bricks. <u>Circle</u> the bricks to show the difference. You can <u>subtract</u> the leftover bricks in any order and quantity. Answers will vary.

15 − 7 = 8

Aha... I circle 8 bricks (the difference), and the 7 bricks that are leftover, I arrange as I want.

15 − ... − ... = 8   15 − ... − ... = 8
15 − ... − ... = 8   15 − ... − ... = 8
15 − ... − ... − ... = 8   15 − ... − ... − ... = 8
15 − ... − ... = 8   15 − ... − ... − ... = 8
15 − ... − ... − ... = 8   15 − ... − ... − ... − ... = 8

2. I have played 12 games in badminton. My brother has won 5 games. <u>How many games</u> have I won?

Played	Brother won	I won
12	5	?7

Won: 5
Played: 12

12 − 5 = 7
Answer: 7 games.

---

1. <u>Cut out</u> 9 equal squares of 3 colors: 3 − pink squares, 3 − green squares, 3 − yellow squares.

So, you have 9 squares of 3 colors.

Now <u>make</u> 1 big square with 9 small squares.

<u>Arrange</u> the squares so that each row, each column, and 1 numeral diagonal of the 9-squared shape has 3 different colors and 1 numeral diagonal has 3 equal colors.

Answers will vary.

2. <u>Follow</u> the directions.
<u>Draw</u> the minute and hour hands red for:
7:00 am. <u>Draw</u> the minute and hour hands black and <u>fill in</u> the missing numbers and letters for: 5 and a half hours later (12:30 pm_____).

3. <u>Write</u> the numbers between:

122 123 124 125   539 540 541 542
483 484 485 486   890 891 892 893
248 249 250 251   965 966 967 968

---

1. <u>Number Sense Strategy</u>. <u>Write</u> subtraction number sentences for each picture. <u>Circle</u> the bricks by 10's. <u>Cross out</u> the number of bricks you subtract. The first one is done for you.

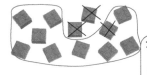

12 − 3 = 9
   ∕ ∖
   2 1   12 − 2 − 1 = 9

I like to count by 10's, so, how to make 10 out of 12? Subtract 2: 12 − 2. And then, subtract the remaining 1.

13 − 9 = 4
   ∕ ∖
   3 6   13 − 3 − 6 = 4

14 − 9 = 5
   ∕ ∖
   4 5   14 − 4 − 5 = 5

12 − 4 = 8
   ∕ ∖
   2 2   12 − 2 − 2 = 8

Page 5

1. <u>Number Sense Strategy.</u> <u>Fill in</u> the missing numbers to complete each number sentence. <u>Use</u> the bricks. The bricks are circled to show the difference. You can <u>subtract</u> the leftover bricks in any order and quantity. *Answers will vary.*

16 - 7 = 9
16 - ... - ... = 9
16 - ... - ... = 9
16 - ... - ... = 9
16 - ... - ... = 9

16 - ... - ... - ... = 9
16 - ... - ... - ... = 9
16 - ... - ... - ... = 9

16 - ... - ... - ... - ... = 9
16 - ... - ... - ... - ... = 9
16 - ... - ... - ... - ... = 9

2. If I do $\frac{2}{8}$ of my homework per day, <u>how much homework</u> will be done in 4 days?

Per day	Days	Homework
$\frac{2}{8}$	4	? $\frac{8}{8}$

$1 = \frac{8}{8}$

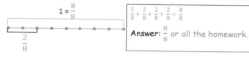

$\frac{2}{8} + \frac{2}{8} + \frac{2}{8} + \frac{2}{8} = \frac{8}{8}$

Answer: $\frac{8}{8}$ or all the homework.

3. <u>Answer</u> the questions.

4972: the sum of the ones and hundreds is 2 + 9 = 11____. The difference between the tens and thousands is 7 - 4 = 3____.

Page 6

1. <u>Answer</u> the questions. <u>Fill in</u> the missing numbers and words.

I had 1 whole apple and a <u>half</u> ($\frac{1}{2}$) of an apple. I shared $\frac{1}{2}$ of an apple with a friend. <u>How many</u> is left?

You can <u>choose</u> one of two strategies:

$1\frac{1}{2} - \frac{1}{2} = \frac{3}{2} - \frac{1}{2} = \frac{3-1}{2} = \frac{2}{2}$ or 1 whole apple_____.

or

$1\frac{1}{2} - \frac{1}{2} = 1$ 1) $\frac{1}{2} - \frac{1}{2} = \frac{1-1}{2} = 0$ 2) 1 - 0 = 1

2. <u>What</u> is the smallest possible 3-digit number if each digit is different? 102.

<u>What</u> is the biggest possible 3-digit number if each digit is different? 987.

3. <u>Find out</u> the numbers hiding behind the bricks. <u>Complete</u> the number sentences.

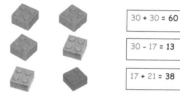

30 + 30 = 60
30 - 17 = 13
17 + 21 = 38

Page 7

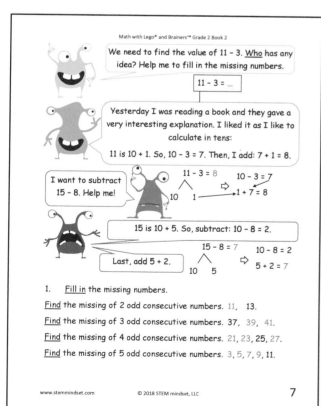

We need to find the value of 11 - 3. <u>Who</u> has any idea? Help me to fill in the missing numbers.

11 - 3 = ...

Yesterday I was reading a book and they gave a very interesting explanation. I liked it as I like to calculate in tens:
11 is 10 + 1. So, 10 - 3 = 7. Then, I add: 7 + 1 = 8.

11 - 3 = 8   10 - 3 = 7
10   1   ⇒ 1 + 7 = 8

I want to subtract 15 - 8. Help me!

15 is 10 + 5. So, subtract: 10 - 8 = 2.

15 - 8 = 7   10 - 8 = 2
10   5   ⇒ 5 + 2 = 7

Last, add 5 + 2.

1. <u>Fill in</u> the missing numbers.

<u>Find</u> the missing of 2 odd consecutive numbers. 11, 13.
<u>Find</u> the missing of 3 odd consecutive numbers. 37, 39, 41.
<u>Find</u> the missing of 4 odd consecutive numbers. 21, 23, 25, 27.
<u>Find</u> the missing of 5 odd consecutive numbers. 3, 5, 7, 9, 11.

Page 8

1. <u>Number Sense Strategy.</u> <u>Write</u> subtraction number sentences for each picture. <u>Circle</u> the bricks by 10's. <u>Cross out</u> the number of bricks you subtract.

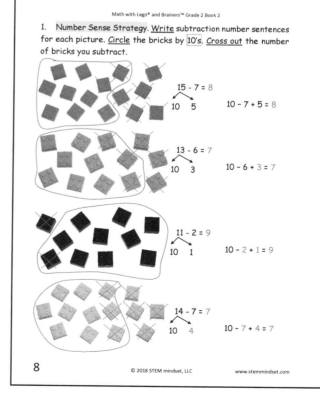

15 - 7 = 8
10   5   10 - 7 + 5 = 8

13 - 6 = 7
10   3   10 - 6 + 3 = 7

11 - 2 = 9
10   1   10 - 2 + 1 = 9

14 - 7 = 7
10   4   10 - 7 + 4 = 7

Page 9

1. Number Sense Strategy. Fill in the missing numbers to complete each number sentence. Use the bricks. Circle the bricks to show the difference. You can subtract the leftover bricks in any order and quantity. *Answers will vary.*

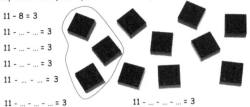

11 - 8 = 3
11 - ... - ... = 3
11 - ... - ... = 3
11 - ... - ... = 3
11 - ... - ... - ... = 3

11 - ... - ... - ... = 3
11 - ... - ... - ... = 3
11 - ... - ... - ... - ... = 3
11 - ... - ... - ... - ... = 3
11 - ... - ... - ... - ... - ... = 3

2. Draw the minute and hour hands red for midnight. How much time is left until 5 past 6 am?

6 hours 5 minutes

Draw the minute and hour hands black.

3. Today is Tuesday, the 12th. If I celebrate my birthday in 13 days, what day will it be (day of the week and date)?

It will be Monday, the 25th.

Page 10

1. We are waiting for 17 guests. I put 9 forks and 5 spoons on the table. How many more forks and spoons do I need?

Guests	Forks	Spoons	More forks	More spoons
17	9	5	? 8	? 12

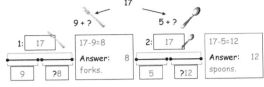

9 + ?   5 + ?

1: 17   17-9=8   Answer: 8 forks.
   9 ?8

2: 17   17-5=12 Answer: 12 spoons.
   5 ?12

2. NSS. Find the value. Do you see a pattern in these problems?

| 13 - 4 + 5 = 14 | 11 - 3 + 4 = 12 |
| 13 - 5 + 4 = 12 | 11 - 4 + 3 = 10 |

| 12 - 4 + 5 = 13 | 12 - 3 + 4 = 13 |
| 12 - 5 + 4 = 11 | 12 - 4 + 3 = 11 |

| 14 - 4 + 5 = 15 | 11 - 4 + 5 = 12 |
| 14 - 5 + 4 = 13 | 11 - 5 + 4 = 10 |

I've noticed that the numbers are the same in each equation in the box (like, 13, 4, and 5), but a different order of addition and subtraction always changes the total by 2; Answers may vary.

Page 11

1. Answer the questions and fill in the missing letters.

Take 3 alike toothpicks.

Break 1 toothpick in the middle into 2 equal parts.

Connect the ends of the 2 broken parts with the 2 whole toothpicks as it's shown in the picture. The 2 opposite sides are equal and parallel.

Sketch your shape in the box.
How many sides does it have?   4 sides.
How many angles does it have?   4 angles.

A shape with 2 pairs of equal (or congruent) sides, which are parallel, and 4 angles is called

p a r a l l e l o g r a m
16 1 18 1 12 12 5 12 15 7 1 13

(to solve this puzzle, look at the back of the book and find out what place each letter takes in the alphabet: A is 1st, Z is 26th).

How many toothpicks do you need to make a parallelogram if one pair of sides has 2 toothpicks each side and another pair of sides has 3 toothpicks each side? 10 toothpicks.

Page 12

1. Number Sense Strategy. Write subtraction number sentences for each picture. Circle the bricks by 10's. Cross out the number of bricks you subtract.

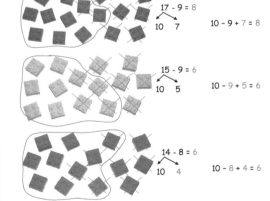

17 - 9 = 8
10 7
10 - 9 + 7 = 8

15 - 9 = 6
10 5
10 - 9 + 5 = 6

14 - 8 = 6
10 4
10 - 8 + 4 = 6

16 - 9 = 7
10 6
10 - 9 + 6 = 7

1. Number Sense Strategy. Fill in the missing numbers to complete each number sentence. Use the bricks. Circle the bricks to show the difference. You can subtract the leftover bricks in any order and quantity. Answers will vary.

12 − 8 = 4
12 − ... − ... = 4
12 − ... − ... = 4
12 − ... − ... = 4
12 − ... − ... = 4

12 − ... − ... − ... = 4   12 − ... − ... − ... = 4
12 − ... − ... − ... = 4   12 − ... − ... − ... − ... = 4
12 − ... − ... − ... − ... = 4   12 − ... − ... − ... − ... − ... = 4

2. I wrote 15, 3, 8, 18, 11, 4 on the blackboard. How many more is the largest number than the smallest number?

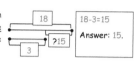

18 − 3 = 15
Answer: 15.

3. Fill in the missing numbers. Find the value. Compare (> or <).

$1 - \frac{5}{6} = \frac{6}{6} - \frac{5}{6} = \frac{6-5}{6} = \frac{1}{6}$ < $1 - \frac{2}{3} = \frac{3}{3} - \frac{2}{3} = \frac{3-2}{3} = \frac{1}{3}$

---

1. Fill in the missing numbers.

Hundreds	Tens	Ones	Number
8	9	4	984
3	7	1	371
7	4	9	749
4	6	2	462
6	3	2	632
8	7	9	879
1	2	6	126
5	0	4	504
2	9	5	295

2. Ask an adult for help to measure 1 pound of beans, peas, apples, and oranges. Find out how many of each fruit or vegetable you have in 1 pound. Answers will vary.

1 pound of beans equals _____ beans.
1 pound of peas equals _____ peas.
1 pound of apples equals _____ apples.
1 pound of oranges equals _____ oranges.

Measure your weight and write it in: ... pounds;
...kilograms.

Measure 1 cup of water with 2 tablespoon of salt:... ounces;
... grams.

Measure 1 cup of pure water: ... ounces;
... grams.

---

1. I drew 5 line segments. Measure the line segments in inches.

A •————————• B   A •————————————• C
A •——————• D   M •——————————————————• N

Add the lengths and write how many inches they make together according to the letters below. Draw the line segments you've got on the grid.

AB + AC = KL = 2 + 2 = 4   AD + AB = NL = 1 + 2 = 3
MN + AD = ML = 3 + 1 = 4   AC + AD = QR = 2 + 1 = 3
AB + MN = PR = 2 + 3 = 5   AC + MN = OP = 2 + 3 = 5

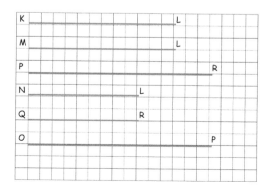

---

1. Number Sense Strategy. Write subtraction number sentences for each picture. Circle the bricks by 10's. Cross out the number of bricks you subtract.

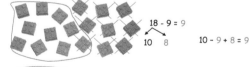
18 − 9 = 9
10   8
10 − 9 + 8 = 9

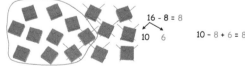
16 − 8 = 8
10   6
10 − 8 + 6 = 8

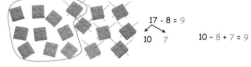
17 − 8 = 9
10   7
10 − 8 + 7 = 9

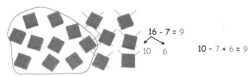
16 − 7 = 9
10   6
10 − 7 + 6 = 9

Page 17

1. **Number Sense Strategy.** <u>Fill in</u> the missing numbers to complete each number sentence. <u>Use</u> the bricks. <u>Circle</u> the bricks to show the difference. You can <u>subtract</u> the leftover bricks in any order and quantity. *Answers will vary.*

13 – 8 = 5

13 – … – … = 5
13 – … – … = 5
13 – … – … = 5
13 – … – … = 5

13 – … – … – … = 5   13 – … – … – … = 5
13 – … – … – … = 5   13 – … – … – … = 5
13 – … – … – … – … = 5 13 – … – … – … – … = 5

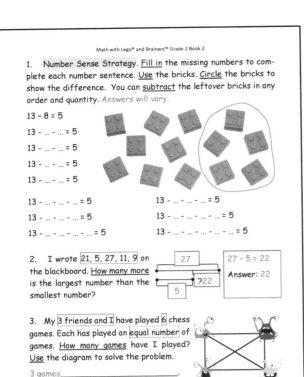

2. I wrote 21, 5, 27, 11, 9 on the blackboard. <u>How many more</u> is the largest number than the smallest number?

   27 – 5 = 22
   Answer: 22

3. My 3 friends and I have played 6 chess games. Each has played an equal number of games. <u>How many games</u> have I played? <u>Use</u> the diagram to solve the problem.

   3 games

Page 18

1. **Number Sense Strategy.** <u>Write</u> subtraction number sentences for each picture. <u>Use</u> the strategy of rounding up the subtrahend.

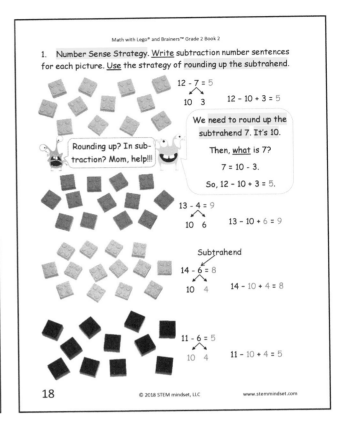

12 – 7 = 5
  /\
10 3   12 – 10 + 3 = 5

We need to round up the subtrahend 7. It's 10. Then, <u>what</u> is 7? 7 = 10 – 3. So, 12 – 10 + 3 = 5.

13 – 4 = 9
  /\
10 6   13 – 10 + 6 = 9

Subtrahend
14 – 6 = 8
  /\
10 4   14 – 10 + 4 = 8

11 – 6 = 5
  /\
10 4   11 – 10 + 4 = 5

Page 19

1. <u>Answer</u> the questions and <u>fill in</u> the missing letters.

Why does one end go up?..
I've used toothpicks!!!

<u>Take</u> 4 alike toothpicks.

<u>Break</u> 1 toothpick in the middle into 2 equal parts and <u>break</u> another toothpick into 2 unequal parts, <u>throw away</u> the smaller part.

<u>Connect</u> the ends of the 3 broken parts with 1 whole toothpick as it's shown in the picture. 2 sides are equal and not parallel. 2 other sides are unequal, but parallel.

<u>Sketch</u> your shape in the box.
<u>How many sides</u> does it have?   4 sides.
<u>How many angles</u> does it have?   4 angles.

A shape with at least 1 pair of parallel sides (which are called the base) and 4 angles is called

t   r   a   p   e   z   o   i   d
20 18 1 16 5   26 15 9   4

(to solve this puzzle, <u>look</u> at the back of the book and <u>find</u> out what place each letter takes in the alphabet: A is 1ˢᵗ, Z is 26ᵗʰ).

Page 20

1. **Number Sense Strategy.** <u>Fill in</u> the missing numbers to complete each number sentence. <u>Use</u> the bricks. <u>Circle</u> the bricks to show the difference. You can <u>subtract</u> the leftover bricks in any order and quantity. *Answers will vary.*

14 – 8 = 6

14 – … – … = 6
14 – … – … = 6
14 – … – … = 6
14 – … – … = 6

14 – … – … – … = 6   14 – … – … – … = 6
14 – … – … – … = 6   14 – … – … – … = 6
14 – … – … – … – … = 6 14 – … – … – … – … = 6

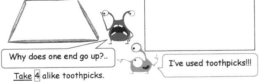

2. <u>Draw</u> the minute and hour hands red for: 8:20 am. My school is over at 1:30pm. <u>How much time</u> do I spend at school?

   3:40 + 1:30 = 5:10 or 5 hours 10 minutes

3. Do you know that any month starting on Sunday, has Friday, the 13ᵗʰ? <u>Check</u> it on the calendar and <u>write</u> the months which have Friday the 13ᵗʰ this year: Answers will vary for each year.

Page 21

1. **Number Sense Strategy.** Write subtraction number sentences for each picture. Use the strategy of rounding up the subtrahend.

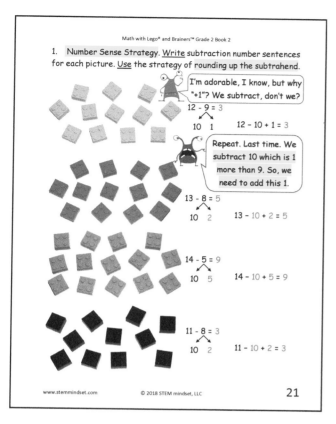

I'm adorable, I know, but why "+1"? We subtract, don't we?

12 − 9 = 3
10   1   12 − 10 + 1 = 3

Repeat. Last time. We subtract 10 which is 1 more than 9. So, we need to add this 1.

13 − 8 = 5
10   2   13 − 10 + 2 = 5

14 − 5 = 9
10   5   14 − 10 + 5 = 9

11 − 8 = 3
10   2   11 − 10 + 2 = 3

Page 22

1. 8 kids came to my birthday party at 2 pm, 6 more kids came at 2:30 pm. 7 kids had to leave at 4:30 pm. How many kids stayed? Solve the problem with two number sentences and then, with one number sentence.

Kids at 2:00	Kids at 2:30	Kids in all	Left at 4:30	Stayed
8	6	? 14	7	? 7

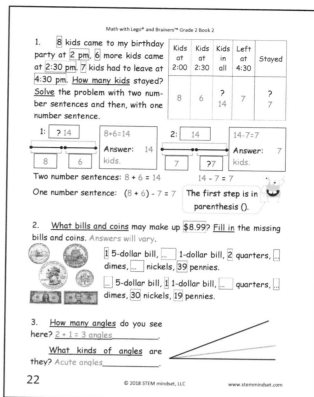

1: ? 14   8+6=14   2: 14   14−7=7
   8   6   Answer: 14 kids.   7   ?7   Answer: 7 kids.

Two number sentences: 8 + 6 = 14   14 − 7 = 7
One number sentence: (8 + 6) − 7 = 7

The first step is in parenthesis ().

2. What bills and coins may make up $8.99? Fill in the missing bills and coins. Answers will vary.

1 5-dollar bill, ... 1-dollar bill, 2 quarters, ... dimes, ... nickels, 39 pennies.

... 5-dollar bill, 1 1-dollar bill, ... quarters, ... dimes, 30 nickels, 19 pennies.

3. How many angles do you see here? 2 + 1 = 3 angles
What kinds of angles are they? Acute angles_____.

Page 23

1. **Number Sense Strategy.** Fill in the missing numbers to complete each number sentence. Use the bricks. Circle the bricks to show the difference. You can subtract the leftover bricks in any order and quantity.

15 − 8 = 7
15 − ... − ... = 7
15 − ... − ... = 7
15 − ... − ... = 7
15 − ... − ... = 7
15 − ... − ... − ... = 7
15 − ... − ... − ... = 7
15 − ... − ... − ... − ... = 7

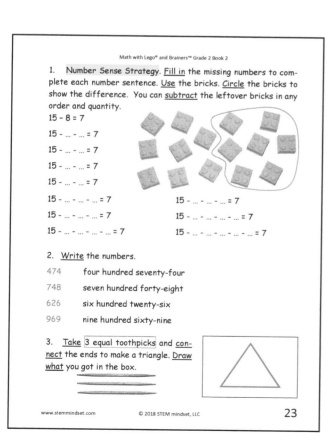

15 − ... − ... = 7
15 − ... − ... − ... = 7
15 − ... − ... − ... − ... = 7

2. Write the numbers.
474   four hundred seventy-four
748   seven hundred forty-eight
626   six hundred twenty-six
969   nine hundred sixty-nine

3. Take 3 equal toothpicks and connect the ends to make a triangle. Draw what you got in the box.

Page 24

1. Fill in the missing numbers and find the value. The first one is done for you.

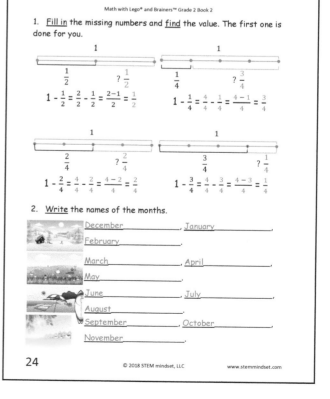

$1 - \frac{1}{2} = \frac{2}{2} - \frac{1}{2} = \frac{2-1}{2} = \frac{1}{2}$

$1 - \frac{1}{4} = \frac{4}{4} - \frac{1}{4} = \frac{4-1}{4} = \frac{3}{4}$

$1 - \frac{2}{4} = \frac{4}{4} - \frac{2}{4} = \frac{4-2}{4} = \frac{2}{4}$

$1 - \frac{3}{4} = \frac{4}{4} - \frac{3}{4} = \frac{4-3}{4} = \frac{1}{4}$

2. Write the names of the months.

December, January,
February,
March, April,
May.
June, July,
August.
September, October,
November.

1. Number Sense Strategy. Write subtraction number sentences for each picture. Use the strategy of rounding up the subtrahend.

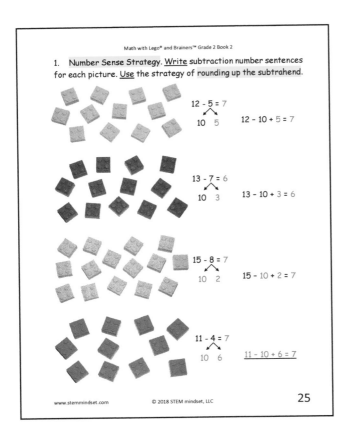

12 - 5 = 7
 10 5   12 – 10 + 5 = 7

13 - 7 = 6
 10 3   13 – 10 + 3 = 6

15 - 8 = 7
 10 2   15 – 10 + 2 = 7

11 - 4 = 7
 10 6   11 – 10 + 6 = 7

---

1. I invited 15 kids to my birthday. My sister put 9 plates (Pl.), 7 spoons (Sp.) and 6 cups(C.) on the table. How many more plates, spoons and cups does she need? Draw the diagrams.

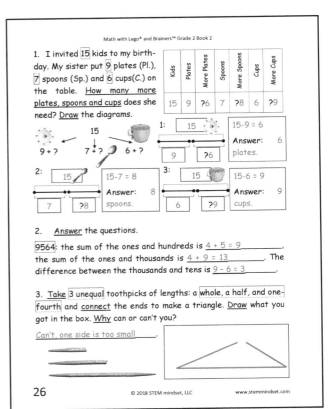

Kids	Plates	More Plates	Spoons	More Spoons	Cups	More Cups
15	9	?6	7	?8	6	?9

1: 15 / 9, ?6   15-9 = 6 Answer: 6 plates.
2: 15 / 7, ?8   15-7 = 8 Answer: 8 spoons.
3: 15 / 6, ?9   15-6 = 9 Answer: 9 cups.

2. Answer the questions.
9564: the sum of the ones and hundreds is 4 + 5 = 9 ___,
the sum of the ones and thousands is 4 + 9 = 13 ___. The difference between the thousands and tens is 9 - 6 = 3 ___.

3. Take 3 unequal toothpicks of lengths: a whole, a half, and one-fourth and connect the ends to make a triangle. Draw what you got in the box. Why can or can't you?

Can't, one side is too small ___.

---

1. Number Sense Strategy. Insert the missing numbers to complete each number sentence. Use the bricks. Circle the bricks to show the difference. You can subtract the leftover bricks in any order and quantity. Answers will vary.

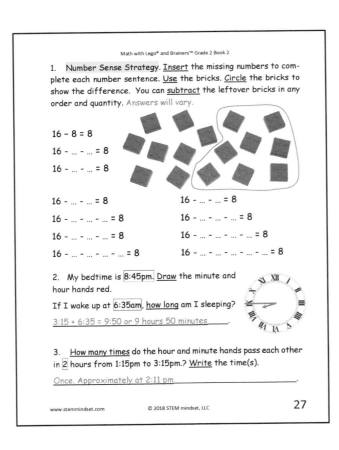

16 – 8 = 8
16 – ... – ... = 8
16 – ... – ... = 8
16 – ... – ... = 8   16 – ... – ... = 8
16 – ... – ... – ... = 8   16 – ... – ... – ... = 8
16 – ... – ... – ... = 8   16 – ... – ... – ... = 8
16 – ... – ... – ... – ... = 8   16 – ... – ... – ... – ... = 8

2. My bedtime is 8:45pm. Draw the minute and hour hands red.
If I wake up at 6:35am, how long am I sleeping?
3:15 + 6:35 = 9:50 or 9 hours 50 minutes ___

3. How many times do the hour and minute hands pass each other in 2 hours from 1:15pm to 3:15pm.? Write the time(s).
Once. Approximately at 2:11 pm ___

---

1. Circle the shapes which are called trapezoids. Write the names of the shapes.

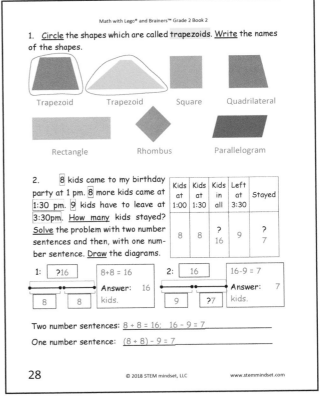

Trapezoid   Trapezoid   Square   Quadrilateral

Rectangle   Rhombus   Parallelogram

2. 8 kids came to my birthday party at 1 pm. 8 more kids came at 1:30 pm. 9 kids have to leave at 3:30pm. How many kids stayed? Solve the problem with two number sentences and then, with one number sentence. Draw the diagrams.

Kids at 1:00	Kids at 1:30	Kids in all	Left at 3:30	Stayed
8	8	? 16	9	? 7

1: ?16 / 8, 8   8+8 = 16 Answer: 16 kids.
2: 16 / 9, ?7   16-9 = 7 Answer: 7 kids.

Two number sentences: 8 + 8 = 16; 16 – 9 = 7 ___
One number sentence: (8 + 8) – 9 = 7 ___

1. **Number Sense Strategy.** <u>Write</u> subtraction number sentences for each picture. <u>Use</u> the strategy of rounding up the subtrahend.

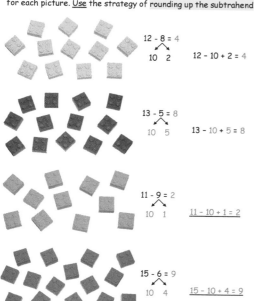

12 − 8 = 4
 10 2   12 − 10 + 2 = 4

13 − 5 = 8
 10 5   13 − 10 + 5 = 8

11 − 9 = 2
 10 1   11 − 10 + 1 = 2

15 − 6 = 9
 10 4   15 − 10 + 4 = 9

---

1. **Number Sense Strategy.** <u>Fill in</u> the missing numbers to complete each number sentence. <u>Use</u> the bricks. <u>Circle</u> the bricks to show the difference. You can <u>subtract</u> the leftover bricks in any order and quantity. *Answers will vary.*

17 − 8 = 9
17 − ... − ... = 9
17 − ... − ... = 9
17 − ... − ... = 9
17 − ... − ... = 9
17 − ... − ... − ... = 9   17 − ... − ... − ... = 9
17 − ... − ... − ... = 9   17 − ... − ... − ... − ... = 9
17 − ... − ... − ... − ... = 9   17 − ... − ... − ... − ... − ... = 9

2. <u>Write</u> the number words and the value of the hundreds, tens, and ones.

458 Four hundred fifty-eight
458 4 hundreds   5 tens   8 ones
341 Three hundred forty-one
341 3 hundreds   4 tens   1 ones
179 One hundred seventy-nine
179 1 hundreds   7 tens   9 ones

---

1. **Number Sense Strategy.** <u>Write</u> subtraction number sentences for each picture. <u>Use</u> the strategy of rounding up the subtrahend.

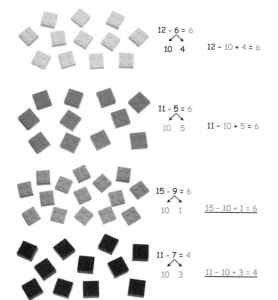

12 − 6 = 6
 10 4   12 − 10 + 4 = 6

11 − 5 = 6
 10 5   11 − 10 + 5 = 6

15 − 9 = 6
 10 1   15 − 10 + 1 = 6

11 − 7 = 4
 10 3   11 − 10 + 3 = 4

---

1. <u>Answer</u> the questions and <u>fill in</u> the missing letters.

<u>Take</u> 3 alike toothpicks and <u>break</u> them in the middle. <u>Connect</u> the ends of <u>all</u> the broken parts as the picture shows.

<u>Sketch</u> your shape in the box.

<u>How many sides</u> does it have?   6 sides.
<u>How many angles</u> does it have?   6 angles.

A closed shape with 6 sides and 6 angles is called a

h e x a g o n
8 5 24 1 7 15 14

(to solve this puzzle, look at the back of the book and find out what place each letter takes in the alphabet: A is 1ˢᵗ, Z is 26ᵗʰ).

2. <u>Look</u> at the picture, <u>answer</u> the question and <u>fill in</u> the numbers.

<u>How many more blue bricks</u> do I have?

14 − 5 = 9 more blue bricks.

1. It's called a Hexagonal Honeycomb. Why did bees choose hexagons? <u>Answers will vary and can include: the six-sided shapes connect together stronger; the shape is more efficient</u>.

2. I have [20] bricks. My brother has [8] bricks. <u>How many more bricks</u> do I have?

I have	Brother has	How many more
20	8	? <u>12 more</u>

Answer: 12 more bricks.

3. <u>Find</u> [3] triangles in the shape and <u>write</u> down the letters, for example, ACD. The order of the letters does not matter.

1. A B C
2. A C D
3. A B D

4. <u>What day</u> is today? <u>Answers will vary</u>.
<u>What day</u> will it be in [9] days? <u>Answers will vary</u>.

---

1. Number Sense Strategy. <u>Fill in</u> the missing numbers to complete each number sentence. <u>Use</u> the bricks. <u>Circle</u> the bricks to show the difference. You can <u>subtract</u> the leftover bricks in any order and quantity. Answers will vary.

18 - 9 = 9
18 - 6 - 3 = 9
18 - ... - ... = 9
18 - ... - ... = 9
18 - ... - ... = 9
18 - ... - ... - ... = 9   18 - ... - ... - ... = 9
18 - ... - ... - ... = 9   18 - ... - ... - ... - ... = 9
18 - ... - ... - ... - ... = 9   18 - ... - ... - ... - ... = 9

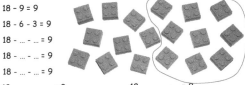

2. We bought [22] pounds of blueberries. We used [8] pounds for making a jam. <u>How many pounds</u> were left?

Bought	Used	Left
22	8	? 14

Bought: 22
Used: 8   Left: 14

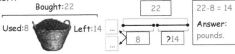

22   22-8 = 14
8   ?14   Answer: 14 pounds.

3. We had Tuesday the day before yesterday. <u>What day of the week</u> will it be after tomorrow? <u>Saturday</u>.

---

1. NSS. <u>Find</u> the value. <u>Do you see</u> a pattern in these problems?

| 14 - 7 + 6 = 13 | 15 - 5 + 6 = 16 |
| 14 - 6 + 7 = 15 | 15 - 6 + 5 = 14 |

| 16 - 7 + 9 = 18 | 13 - 8 + 4 = 9 |
| 16 - 9 + 7 = 14 | 13 - 4 + 8 = 17 |

I've noticed that <u>the numbers are the same in each equation in the box (like, 14, 7, and 6), but a different order of addition and subtraction changes the total; Answers may vary</u>.

2. <u>Write</u> the numbers between:

367 368 369 370   134 135 136 137
895 896 897 898   471 472 473 474
546 547 548 549   621 622 623 624

3. <u>Look</u> at the picture below. <u>Take</u> [12] toothpicks to <u>make</u> the same shape. <u>Take</u> [2] toothpicks <u>away</u> so that you have [2] unequal squares. <u>Draw</u> your shape in the box.

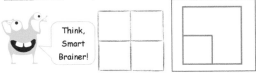

Think, Smart Brainer!

---

1. I solved [23] problems. My friend solved [5] problems less. <u>How many problems</u> did my friend solve?

I solved	Friend solved	Friend solved
23	5 <u>less</u>	? 18

I: 23
My friend: 18

23   23-5 = 18
?18   5   Answer: 18 problems.

2. NSS. <u>Find</u> the value. <u>Do you see</u> a pattern in these problems?

| 13 - 9 + 8 = 12 | 14 - 7 + 9 = 16 |
| 13 - 8 + 9 = 14 | 14 - 9 + 7 = 12 |

| 15 - 6 + 9 = 18 | 16 - 8 + 4 = 12 |
| 15 - 9 + 6 = 12 | 16 - 4 + 8 = 20 |

I've noticed that <u>the numbers are the same in each equation in the box,), but a different order of addition and subtraction changes the total; Answers may vary</u>.

3. <u>Let's</u> guess the dimensions of your table, and then, <u>measure</u> it. Answers will vary.

<u>How long</u> is your table?

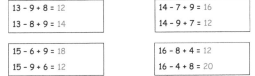

Guess: ...
True: ...

Page 37

1. I helped Mom to bake 24 chocolate chip cookies and 9 sprinkle cookies. Which cookie did we bake more of? How many more?

Choco-late	Sprin-kle	How many more
24	9	? 15 more

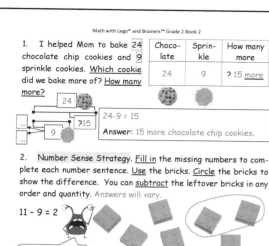

24 - 9 = 15

Answer: 15 more chocolate chip cookies.

2. Number Sense Strategy. Fill in the missing numbers to complete each number sentence. Use the bricks. Circle the bricks to show the difference. You can subtract the leftover bricks in any order and quantity. Answers will vary.

11 - 9 = 2

Ok, I will help you with...the first one 😊.

11 - 1 - 8 = 2
11 - ... - ... = 2
11 - ... - ... - ... = 2
11 - ... - ... - ... - ... = 2
11 - ... - ... - ... - ... - ... = 2

11 - ... - ... = 2
11 - ... - ... = 2
11 - ... - ... - ... = 2
11 - ... - ... - ... - ... = 2
11 - ... - ... - ... - ... - ... - ... = 2

Page 38

1. I jumped 21 times on the trampoline. My sister jumped 9 times. Who jumped more? How many more? Draw the diagram.

I jumped	Sister jumped	How many more
21	9	? 12 more

I: 21   Sis: 9

21 - 9 = 12

Answer: I jumped 12 times more.

2. Number Sense Strategy. Fill in the missing numbers to complete each number sentence. Use the bricks. Circle the bricks to show the difference. You can subtract the leftover bricks in any order and quantity. Answers will vary.

14 - 9 = 5
14 - ... - ... = 5
14 - ... - ... = 5
14 - ... - ... = 5
14 - ... - ... = 5
14 - ... - ... - ... = 5
14 - ... - ... - ... = 5
14 - ... - ... - ... - ... = 5

14 - ... - ... - ... = 5
14 - ... - ... - ... - ... = 5
14 - ... - ... - ... - ... - ... = 5

3. My brother and sister are 11 years old altogether. Brother is 3 years older than my sister. How old are they? 7 and 4 years old

Page 39

1. I need 23 invitation cards for my birthday party. I have made 7 cards. How many cards are left to be made? Fill in the table.

Need	Made	Left
23	7	? 16

Need: 23
Made: 7   Left: 16

23 - 7 = 16

Answer: 16 cards.

2. Find the value. Compare the fractions and circle >, <, or =.

$\frac{1}{2}$ (>) < = $\frac{3}{8}$

$\frac{2}{4}$ > < (=) $\frac{4}{8}$

 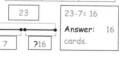

$\frac{1}{2} + \frac{2}{2} = \frac{1+2}{2} = \frac{3}{2}$ (>) < = $\frac{2}{4} + \frac{2}{4} = \frac{2+2}{4} = \frac{4}{4}$

$\frac{5}{8} + \frac{2}{8} = \frac{5+2}{8} = \frac{7}{8}$ > (<) = $\frac{3}{4} + \frac{1}{4} = \frac{3+1}{4} = \frac{4}{4}$

3. Let's guess the dimensions of your table, and then, measure it (use inches or centimeters). Answers will vary.

How wide is your table?   Guess: ......   True: ......
How high is your table?   Guess: ......   True: ......

Page 40

1. I made 15 Valentine's cards for kids from my class and 9 less cards for kids from the swim team. How many cards did I make for kids from the swim team?

Class	Swim team	Swim team
15	9 less	? 6

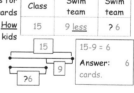

15 - 9 = 6

Answer: 6 cards.

2. Number Sense Strategy. Fill in the missing numbers to complete each number sentence. Use the bricks. Circle the bricks to show the difference. You can subtract the leftover bricks in any order and quantity. Answers will vary.

17 - 9 = 8
17 - ... - ... = 8
17 - ... - ... = 8
17 - ... - ... = 8
17 - ... - ... = 8
17 - ... - ... - ... = 8
17 - ... - ... - ... = 8
17 - ... - ... - ... - ... = 8

17 - ... - ... - ... = 8
17 - ... - ... - ... - ... = 8
17 - ... - ... - ... - ... - ... = 8

3. My sister is 5 years older than me. The sum of our ages equals 17 years. How old are we? 11 and 6 years old   .

Page 41

1. Last weekend I was fishing on a farm. I caught 25 fish, 6 of them were small. We put them back. How many fish did I bring home? Draw the diagram.

Caught	Let back	Brought home
25	6	? 19

Caught:25
Let back:6 Brought:19

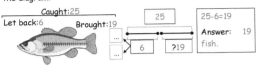

25-6=19
Answer: 19 fish.

2. Number Sense Strategy. Fill in the missing numbers to complete each number sentence. Use the bricks. Circle the bricks to show the difference. You can subtract the leftover bricks in any order and quantity.
Answers will vary.

13 - 9 = 4

13 - 3 - 6 = 4
13 - ... - ... = 4
13 - ... - ... - ... = 4
13 - ... - ... - ... = 4
13 - ... - ... - ... - ... = 4

13 - ... - ... = 4
13 - ... - ... = 4
13 - ... - ... - ... = 4
13 - ... - ... - ... - ... = 4
13 - ... - ... - ... - ... - ... = 4

Page 42

1. NSS. Find the value.

13 - 7 + 8 = 14
13 - 8 + 7 = 12

18 - 9 + 6 = 15
18 - 6 + 9 = 21

15 - 8 + 3 = 10
15 - 3 + 8 = 20

11 - 2 + 9 = 18
11 - 9 + 2 = 4

How do we calculate addition and subtraction in one equation?
We calculate addition or subtraction ONLY from LEFT to Right:
first, 13 - 7, then, 6 + 8. Or first, 13-8, then, 5 + 7 .

2. Our neighbor picked up 14 pounds of honey. He sold 9 pounds. How many pounds were left over? Fill in the table.

Picked up	Sold out	Left
14	9	? 5

Picked:14
Sold:9 Left:5

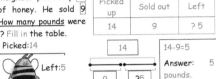

14-9=5
Answer: 5 pounds.

3. Find the value.

$\frac{1}{5} + \frac{1}{5} = \frac{2}{5}$ $\frac{1}{5} + \frac{4}{5} = \frac{5}{5} = 1$ $\frac{2}{5} + \frac{3}{5} = \frac{5}{5} = 1$

$\frac{3}{5} - \frac{1}{5} = \frac{2}{5}$ $\frac{5}{5} - \frac{4}{5} = \frac{1}{5}$ $\frac{4}{5} - \frac{1}{5} = \frac{3}{5}$

Page 43

1. It takes us 15 minutes to go to school. We have been driving for 7 minutes. How much more time is left? Fill in the table and draw the diagram.

Time to school	Have been driving	Left
15	7	? 8

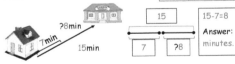
?8min
7min 15min

15
7 ?8
15-7=8
Answer: 8 minutes.

2. Number Sense Strategy. Fill in the numbers to complete each number sentence. Circle the bricks to show the difference. You can subtract the leftover bricks in any order and quantity. Answers will vary.

12 - 9 = 3

12 - 8 - 1 = 3
12 - ... - ... = 3
12 - ... - ... - ... = 3
12 - ... - ... - ... = 3
12 - ... - ... - ... - ... = 3

12 - ... - ... = 3
12 - ... - ... = 3
12 - ... - ... - ... = 3
12 - ... - ... - ... - ... = 3
12 - ... - ... - ... - ... - ... = 3

Page 44

1. I scored 15 balls and missed 5 more balls than I scored. How many balls did I miss? Draw the diagram.

Hit	Missed	Missed
15	5 more	? 20

?20
15
15+5 = 20
Answer: 20 balls.

2. I played 15 soccer games and 5 basketball games this season. How many less basketball games did I play? Draw the diagram.

Soccer	Basketball	Basketball
15	5	? 10 less

15 10 less

15
5 ?10
15-5 = 10
Answer: 10 less.

3. Fill in the missing numbers.

Hundreds	Tens	Ones	Number
6	2	7	627
9	5	3	953
2	1	8	218
5	8	6	586
1	3	5	135
7	2	1	721
4	6	7	467
9	5	3	953
3	4	8	348

Page 45

1. <u>Answer</u> the questions, using a calendar. *Answers will vary.*

<u>How many days</u> do you study in December?   ... days.
<u>How many days</u> are weekends?   ... days.
<u>How many days</u> are you on vacation?   ...days.
<u>How many days</u> do you study in January?   ... days.
<u>How many days</u> are weekends?   ... days.
<u>How many days</u> are you on vacation?   ...days.
<u>How many days</u> do you study in February?   ... days.
<u>How many days</u> are weekends?   ... days.
<u>What month</u> has the most weekends?   _____
<u>What month</u> do you study the most?   _____
<u>What month</u> is your favorite?   _____
<u>Fill in</u> the column graph.

Winter

31
26
21
16
11
6
1
   December   January   February
   ■ Days of vacation   ■ Weekends   ■ Days to study

Page 46

1. <u>Fill in</u> the missing numbers in the Subtraction table. <u>Subtract</u> the number in a row from the number in a column. Some of them are done for you. If you see "-", it means you cannot find the difference (yet!). <u>Color</u> the boxes with the even numbers blue.

_	1	2	3	4	5	6	7	8	9	10	11	12	13	14	15	16	17	18
1	0	1	2	3	4	5	6	7	8	9	10	11	12	13	14	15	16	17
2	-	0	1	2	3	4	5	6	7	8	9	10	12	12	13	14	15	16
3	-	-	0	1	2	3	4	5	6	7	8	9	10	11	12	13	14	15
4	-	-	-	0	1	2	3	4	5	6	7	8	9	10	11	12	13	14
5	-	-	-	-	0	1	2	3	4	5	6	7	8	9	10	11	12	13
6	-	-	-	-	-	0	1	2	3	4	5	6	7	8	9	10	11	12
7	-	-	-	-	-	-	0	1	2	3	4	5	6	7	8	9	10	11
8	-	-	-	-	-	-	-	0	1	2	3	4	5	6	7	8	9	10
9	-	-	-	-	-	-	-	-	0	1	2	3	4	5	6	7	8	9

2. <u>Find</u> the value. <u>Compare</u> the fractions (<u>use</u> > or <).

$\frac{5}{5} - \frac{1}{5} = \frac{4}{5}$ > $\frac{4}{5} - \frac{1}{5} = \frac{3}{5}$ > $\frac{3}{5} - \frac{2}{5} = \frac{1}{5}$

$\frac{3}{5} + \frac{1}{5} - \frac{2}{5} = \frac{2}{5}$ < $\frac{4}{5} - \frac{2}{5} + \frac{3}{5} = \frac{5}{5}$ > $\frac{2}{5} + \frac{3}{5} - \frac{4}{5} = \frac{1}{5}$

$1/5$

Page 47

1. <u>Color</u> the boxes with the black odd numbers green.

_	1	2	3	4	5	6	7	8	9	10	11	12	13	14	15	16	17	18
1	0	1	2	3	4	5	6	7	8	9	10	11	12	13	14	15	16	17
2	-	0	1	2	3	4	5	6	7	8	9	10	11	12	13	14	15	16
3	-	-	0	1	2	3	4	5	6	7	8	9	10	11	12	13	14	15
4	-	-	-	0	1	2	3	4	5	6	7	8	9	10	11	12	13	14
5	-	-	-	-	0	1	2	3	4	5	6	7	8	9	10	11	12	13
6	-	-	-	-	-	0	1	2	3	4	5	6	7	8	9	10	11	12
7	-	-	-	-	-	-	0	1	2	3	4	5	6	7	8	9	10	11
8	-	-	-	-	-	-	-	0	1	2	3	4	5	6	7	8	9	10
9	-	-	-	-	-	-	-	-	0	1	2	3	4	5	6	7	8	9

2. <u>Answer</u> the questions.

I have 2 sticks below. <u>How</u> can I make them equal?

Measure both sticks and cut off the difference from the longest stick_____.

If I cut a stick, <u>should</u> it be the longest or the shortest stick?
The longest stick_____.

<u>How many</u> inches should be cut? 1 inch_____.

Page 48

1. <u>Complete</u> each number sentence. <u>Try</u> to make 10's as it's easier to calculate them. You can <u>draw</u> the arrows to help you make 10's.

17 - 3 - 1 - 6 - 1 = 6   13 - 8 - 2 - 1 - 1 = 1
20 - 4 - 5 - 4 - 5 = 2   20 - 5 - 7 - 4 - 2 = 2
15 + 4 - 13 + 6 = 12   14 - 7 + 10 - 9 = 8
12 + 5 - 11 + 4 = 10   11 - 7 + 15 - 12 = 7

2. <u>Look</u> at the shape below. <u>Take</u> 14 toothpicks and <u>make</u> the same shape. <u>Take</u> 3 toothpicks <u>away</u> so that you will be left with 3 equal squares. <u>Draw</u> the shape you got in the box.

3. <u>Let's</u> guess the dimensions of your chair, and then, <u>measure</u> it (use inches or centimeters). *Answers will vary.*

<u>How high</u> is your chair?   Guess: ...____   True: ...____
<u>How high</u> is your seat?   Guess: ...____   True: ...____

Height of your chair   Height of your seat

1. <u>Use</u> books, <u>Google</u>, or <u>ask</u> an adult to answer these questions.

The longest day of the year 2018 is June the 21st (called June or Summer Solstice). The shortest day of the year 2018 is December the 21st (called Winter Solstice). The days grow longer from December and grow shorter from June.

<u>When</u> are the days longer:
- in January or in February?   February
- in February or in March?   March
- in March or in May?   May

<u>When</u> are the days shorter:
- in July or in September?   September
- in October or in November?   November
- in December or in August?   December

There are 2 days of the year when the day almost equals the night. It's called an equinox. It comes around March the 20th or 21st and September the 22nd or 23rd. On these days <u>how long</u> is:
- the night?   Around 12 hours
- the day?   Around 12 hours

September the 22nd or 23rd is the first day of Fall, and March the 20th or 21st is the first day of Spring.

<u>Do</u> you know any other days that tell us about the start of any season?   The first day of June (summer) or December (winter). Answers will vary _____ .

"I had 7 bricks. I added more and now I have 13 bricks. How many bricks did I add?" Don't you think, guys, this problem is too co-mpli-ca-ted... I don't get it ☺. I'm bad at math☹.

Wow! You do not understand a word, do you?! What will I do, then? I know, I should quit math! ☹.

Calm down, guys ☺! Nobody is going to quit! Let's do it together.

I will help! "I had 7 bricks." Take 7 green bricks. It says, "I added more." If you do not know the number, name it "X" or put "?": 7 + X. Then, you need to add as many bricks as you need to have 13 in all. Add blue bricks until you have 13 bricks altogether: 7 + X = 13. Do you have 13 bricks? Great! Now, count how many blue bricks you added.

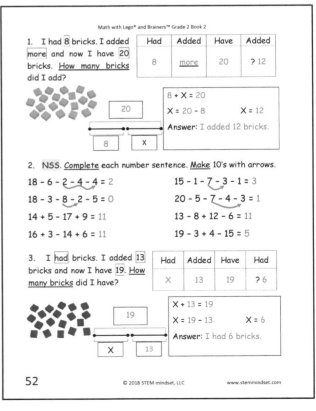

1. I had 8 bricks. I added more and now I have 20 bricks. How many bricks did I add?

Had	Added	Have	Added
8	more	20	? 12

8 + X = 20
X = 20 - 8   X = 12
Answer: I added 12 bricks.

2. NSS. <u>Complete</u> each number sentence. <u>Make</u> 10's with arrows.

18 - 6 - 2 - 4 - 4 = 2   15 - 1 - 7 - 3 - 1 = 3
18 - 3 - 8 - 2 - 5 = 0   20 - 5 - 7 - 4 - 3 = 1
14 + 5 - 17 + 9 = 11   13 - 8 + 12 - 6 = 11
16 + 3 - 14 + 6 = 11   19 - 3 + 4 - 15 = 5

3. I had bricks. I added 13 bricks and now I have 19. How many bricks did I have?

Had	Added	Have	Had
X	13	19	? 6

X + 13 = 19
X = 19 - 13   X = 6
Answer: I had 6 bricks.

1. Number Sense Strategy. Fill in the missing numbers to complete each number sentence. Use the bricks. Circle the bricks to show the difference. You can subtract the leftover bricks in any order and quantity (e.g. = for example). Answers will vary.

16 − 9 = 7   16 − ... − ... = 7

Did you notice a trick with 9? Look, 7 + ... = 16: the ones in the sum is 1 less than the addend 7, right? Or 3 + ... = 12. Yes, 3 + 9 = 12.

Aha, the second addend should be 9, as 9 needs 1 more to make 10: 7 − 1 = 6 (sum is 16), 3 − 1 = 2 (sum is 12).

16 − ... − ... = 7
16 − ... − ... = 7
16 − ... − ... = 7
16 − ... − ... − ... = 7

16 − ... − ... − ... = 7
16 − ... − ... − ... = 7
16 − ... − ... − ... − ... = 7

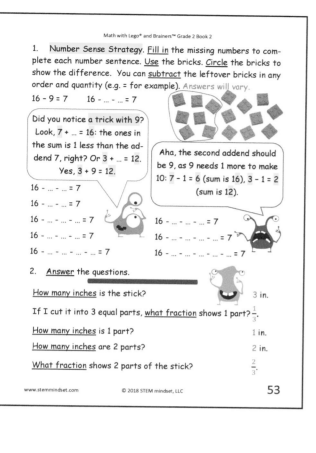

2. Answer the questions.

How many inches is the stick?   3 in.

If I cut it into 3 equal parts, what fraction shows 1 part? $\frac{1}{3}$.

How many inches is 1 part?   1 in.

How many inches are 2 parts?   2 in.

What fraction shows 2 parts of the stick? $\frac{2}{3}$.

---

1. I got a set with 85 bricks. My little sister got a set with 45 bricks less. How many bricks were there in both sets?

	I had	Sister had	Sister had	Both had
	85	45 less	? 40	? 125

1: 85 / ?40 / 45   85−45=40   Answer: 40
2: ?125 / 85 / 40   85+40=125   Answer: 125

Write one number sentence for the problem: 85 + (85 − 45) = 125

Aha, the first step is in parenthesis.

2. Complete each number sentence.

Fine☺. I calculate from the left to the right: 17−4=13, write 13; 13−10=3, write 3; 3+9=12, write 12; 12+5=17. Done!

17 − 4 − 10 + 9 + 5 = 17
   13   3   12

15 − 8 − 3 + 14 − 7 = 11
    7   4   18

3. Let's guess the dimensions of your book, and then, measure it (use inches or centimeters). Answers will vary.

How long is your book?   Guess: .....   True: .....
How wide is your book?   Guess: .....   True: .....

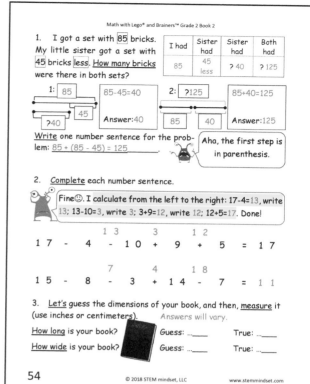

---

1. I had 14 candies. Yesterday I ate 5 of them. How many candies are left?

Had	Ate	Left
14	5	? 9

14 / 5 / ?9   14−5=9   Answer: 9 candies.

2. I had 14 candies. My brother had 5 candies less. How many candies did he have?

I had	Brother	Brother
14	5 less	? 9

14 / ?9 / 5   14−5=9   Answer: 9 candies.

3. Look at the picture below. Take 24 toothpicks and make 6 squares. Take away 2 toothpicks and make 7 equal squares with the remaining toothpicks. Draw the shape or shapes you got in the box.

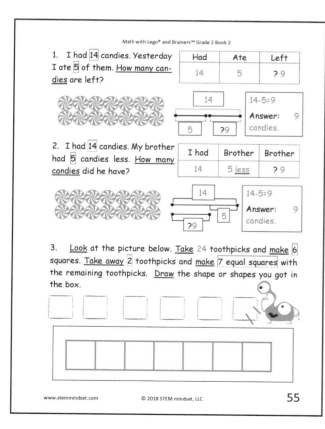

---

1. Cut out 16 equal squares of 4 colors: 4 − orange squares, 4 − red squares, 4 − green squares, 4 − blue squares.

So, you have 16 squares of 4 colors. Draw 1 big square divided into 16 small squares.

Arrange the squares so that each row, column, and numeral diagonal has 4 different colors.

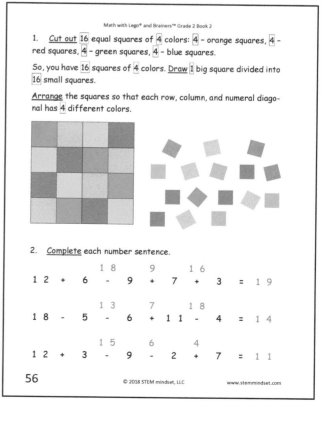

2. Complete each number sentence.

12 + 6 − 9 + 7 + 3 = 19
    18   9   16

18 − 5 − 6 + 11 − 4 = 14
    13   7   18

12 + 3 − 9 − 2 + 7 = 11
    15   6   4

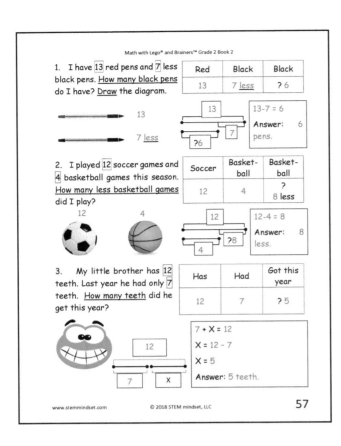

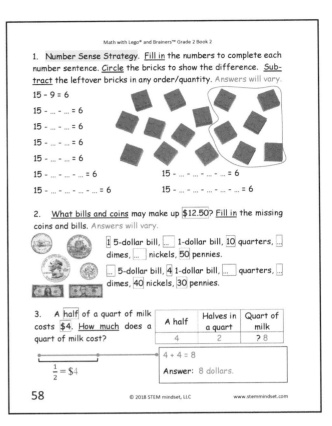

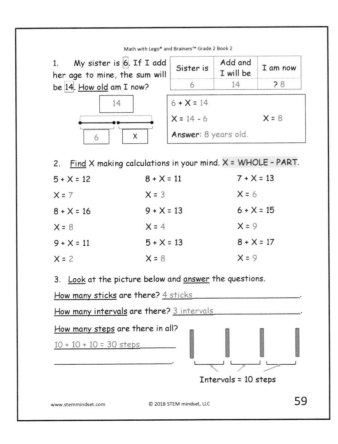

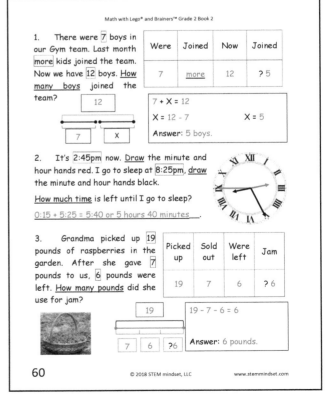

Page 61

1. <u>Find</u> a number behind a creature, making calculations in your mind.

17 - 🐛 = 9   13 - 🐌 = 5   15 - 🐚 = 6
🐛 = 8   🐌 = 8   🐚 = 9

13 - 🐞 = 8   11 - ➤ = 2   16 - 🐁 = 9
🐞 = 5   ➤ = 9   🐁 = 7

12 - 🐀 = 6   14 - 🌸 = 5   11 - 🐛 = 8
🐀 = 6   🌸 = 9   🐛 = 3

2. <u>How many angles</u> do you see here? <u>5 + 4 + 3 + 2 + 1 = 15 angles</u>.
<u>How many</u> are obtuse angles? 3____.
<u>How many</u> are acute angles? 12____.

Look, I explore angles in the black-black lines and mark them orange: 1 + 1 + 1 + 1 + 1 = 5. Then, you will take blue-black lines, then, red-black lines, then, green-black lines, at last, gold-black lines. Done!

I will take blue-black lines: blue-red + blue-green + blue-gold + blue-black. 1+1+1+1=4.

Page 62

1. I found 29 4-stud bricks and 11 more 6-stud bricks. <u>How many bricks</u> did I find in all?

	4-stud	6-stud	6-stud	In all
	29	11 more	? 40	? 69

1: [? 40]   29 + 11 = 40   Answer: 40 bricks.
    [29] [11]

2: [? 69]   29 + 40 = 69   Answer: 69 bricks.
    [29] [40]

<u>Write</u> one number sentence for the problem:
<u>29 + (29 + 11)</u>_____.

2. I poured 10 <u>halves</u> of a cup. <u>How many cups</u> are now in the pan?

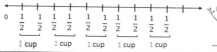

0   $\frac{1}{2}$ $\frac{1}{2}$ $\frac{1}{2}$ $\frac{1}{2}$ $\frac{1}{2}$ $\frac{1}{2}$ $\frac{1}{2}$ $\frac{1}{2}$ $\frac{1}{2}$ $\frac{1}{2}$
  1 cup   1 cup   1 cup   1 cup   1 cup

1 cup = 2 halves of a cup. So, 1 cup = $\frac{1}{2} + \frac{1}{2} = \frac{1+1}{2} = \frac{2}{2} = 1$

1 + 1 + 1 + 1 + 1 = 5
Answer: <u>5 cups</u>_____

3. <u>Cross out</u> the picture which is different.

Page 63

1. The sum of 3 numbers is 20. The sum of the first and second numbers is 12. The difference of the third and second numbers is 1. <u>Find</u> the numbers.
<u>5 + 7 + 8 = 20; 5 + 7 = 12; 8 - 7 = 1</u>_____.

2. <u>Insert</u> the missing numbers hiding behind my pets.

11 + 11 + 11 + 11 = 44
11 + 10 + 10 = 31
11 + 10 - 5 = 16

3. <u>Complete</u> the number sentence.

            1 3   3   1 2
1 7 - 4 - 1 0 + 9 + 5   =   1 7

4. <u>Write in</u> the missing numbers to balance my toys.

4 owls = 12 frogs   3 frogs = ? 1 owl(s)

Page 64

1. For December <u>let's make</u> a project "Daytime" ☺.

You need to <u>check</u> the day length every day and <u>write</u> the hours and minutes of sunrise and sunset in the table. You can <u>use</u> websites Solartopo.com, Timeanddate.com, Aa.usno.navy.mil or you can use Google.com. Length - is the difference between the sunset and the sun rise. Dif. – is the difference between the lengths of 2 days. It's "+" if the daytime goes up, and it's "-" if the daytime goes down. The Dif. Is unknown for the 1st of December.

Date	Day length			Dif.	Date	Day length			Dif.
	Srise	Sset	Length			Srise	Sset	Length	
1					16				
2					17				
3					18				
4					19				
5					20				
6					21				
7					22				
8					23				
9					24				
10					25				
11					26				
12					27				
13					28				
14					29				
15					30				
					31				

<u>Add</u> all the differences for the day length for the whole month:
<u>Answers will vary</u>_____.

1. <u>Make</u> two Line graphs, one for Sunrise and one for Sunset time in December in your area according to your table.

A line Graph shows how the data changes continuously over time.

<u>Use</u> dots to show the time and line segments to connect the dots and show how the time changes.

The dots and line segments for Sunrise are blue and the dots and line segments for Sunset are red. *Answers will vary.*

Sunrise in December

(Sunrise Time)

1 2 3 4 5 6 7 8 9 10 11 12 13 14 15 16 17 18 19 20 21 22 23 24 25 26 27 28 29 30 31 — Srise

---

Sunset in December

(Sunset Time)

1 2 3 4 5 6 7 8 9 10 11 12 13 14 15 16 17 18 19 20 21 22 23 24 25 26 27 28 29 30 31 — Sset

2. ⎡3⎤ Peaches and ⎡3⎤ oranges are in balance with ⎡2⎤ oranges and ⎡4⎤ peaches. <u>Which fruit</u> is heavier: a peach or an orange? <u>They are equal</u>

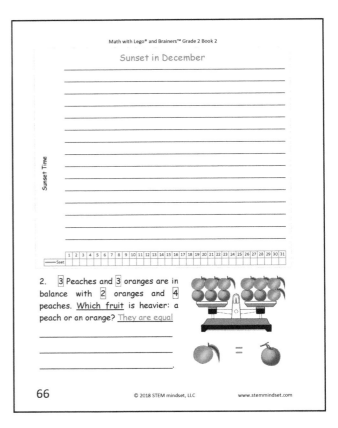

_____

---

1. <u>Compare</u> the fractions. <u>Use</u> "=", ">" or "<" and the shape below.

$\frac{3}{4} > \frac{3}{8}$ $\frac{2}{8} > \frac{2}{12}$

$\frac{1}{4} = \frac{2}{8}$ $\frac{2}{4} > \frac{4}{12}$

$\frac{9}{12} = \frac{3}{4}$ $\frac{2}{8} = \frac{3}{12}$

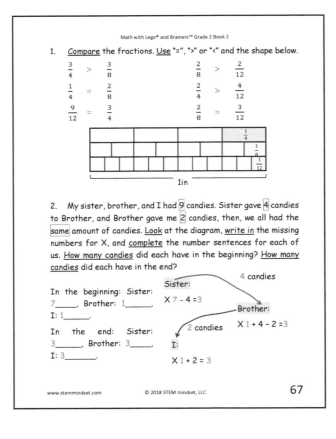

2. My sister, brother, and I had ⎡9⎤ candies. Sister gave ⎡4⎤ candies to Brother, and Brother gave me ⎡2⎤ candies, then, we all had the ⎡same⎤ amount of candies. <u>Look</u> at the diagram, <u>write in</u> the missing numbers for X, and <u>complete</u> the number sentences for each of us. <u>How many candies</u> did each have in the beginning? <u>How many candies</u> did each have in the end?

In the beginning: Sister: 7____, Brother: 1____, I: 1____.

In the end: Sister: 3____, Brother: 3____, I: 3____.

Sister: X 7 - 4 = 3
4 candies →
Brother: X 1 + 4 - 2 = 3
2 candies →
I: X 1 + 2 = 3

---

1. <u>Answer</u> the questions and <u>fill in</u> the missing letters.

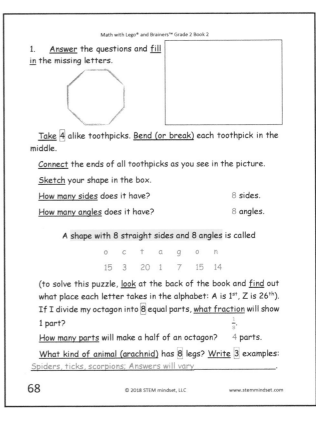

<u>Take</u> ⎡4⎤ alike toothpicks. <u>Bend (or break)</u> each toothpick in the middle.

<u>Connect</u> the ends of all toothpicks as you see in the picture.

<u>Sketch</u> your shape in the box.

<u>How many sides</u> does it have?   8 sides.

<u>How many angles</u> does it have?   8 angles.

A shape with 8 straight sides and 8 angles is called

o c t a g o n
15 3 20 1 7 15 14

(to solve this puzzle, <u>look</u> at the back of the book and <u>find</u> out what place each letter takes in the alphabet: A is 1st, Z is 26th).

If I divide my octagon into ⎡8⎤ equal parts, <u>what fraction</u> will show 1 part? $\frac{1}{8}$.

<u>How many parts</u> will make a half of an octagon?   4 parts.

<u>What kind of animal</u> (arachnid) has ⎡8⎤ legs? <u>Write</u> ⎡3⎤ examples:

<u>Spiders, ticks, scorpions; Answers will vary</u>_____.

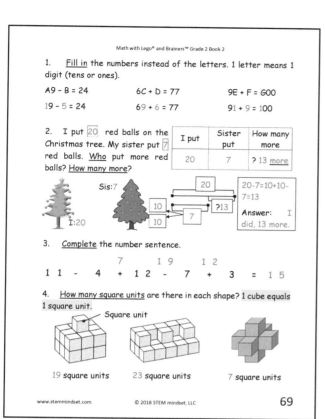

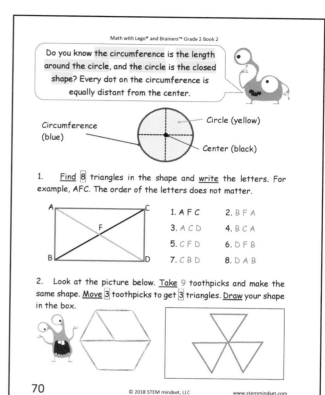

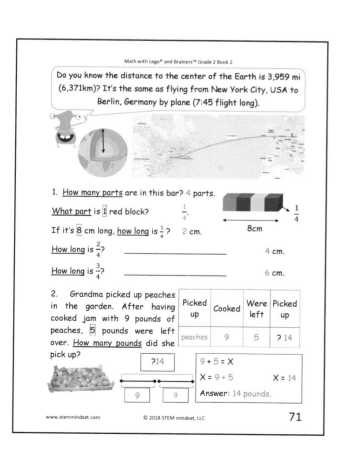

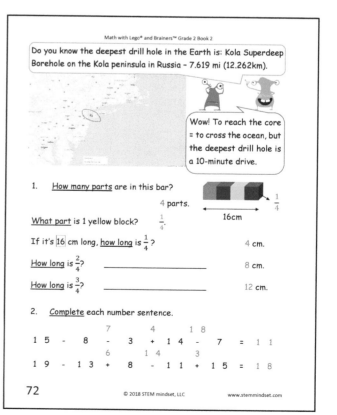

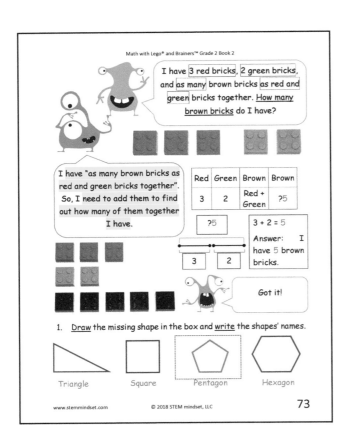

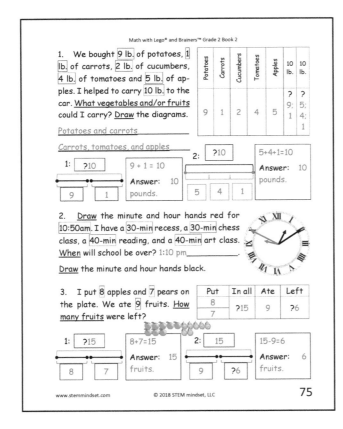

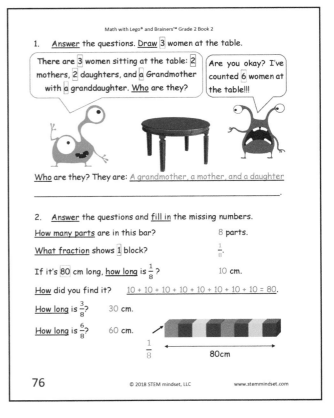

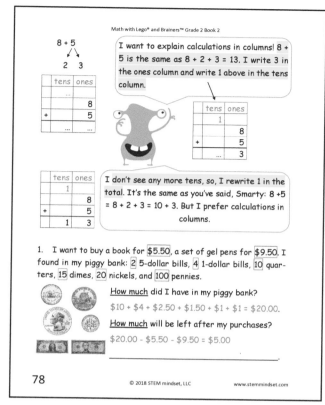
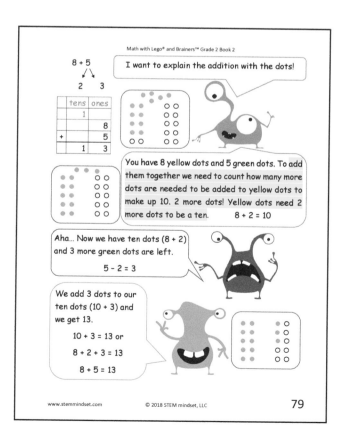

1. Find X by mental math. Remember: X(Part1) = Whole - Part2.

X + 5 = 11	X + 7 = 14	X + 6 = 15
X = 6	X = 7	X = 9
X + 8 = 12	X + 9 = 14	X + 5 = 13
X = 4	X = 5	X = 8
X + 9 = 11	X + 7 = 12	X + 8 = 11
X = 2	X = 5	X = 3

3. Continue the pattern: 100, 95, 85, 80, 70, 65, 55, 50.

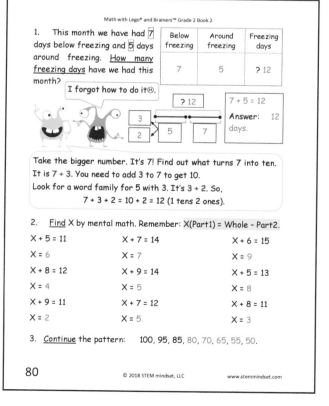

1. Draw 3 unequal quadrilaterals with a ruler. Cut each one into 4 triangles. How many triangles did you get in each quadrilateral?
8 triangles, 8 triangles, 8 triangles.

Draw the shapes of your biggest and smallest triangles in the box.

Shapes will vary.
The biggest triangle: The smallest triangle:

2. I had 15 chocolate cookies and 8 butter cookies. I gave 6 chocolate cookies to my sister and she gave me 7 butter cookies. How many chocolate cookies do I have now? How many butter cookies do I have now?

I - ch.	I - but.	I → Sis	I → Me	I - ch.	I - but.
15	8	6	7	?9	?15

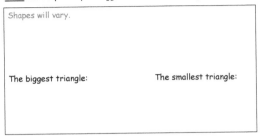

1:	15		15-6=9		2:	?15		8+7=15
6	?9		Answer: 9 cookies.		8	7		Answer: 15 cookies.

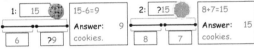

---

1. Draw the minute and hour hands red for 10am.

How much time is left until 3:55 pm?
5 hours 55 minutes

Draw the minute and hour hands black.

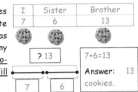

2. I have 7 chocolate cookies and my sister has 6 chocolate cookies. My brother has as many chocolate cookies as my sister and me. How many chocolate cookies does he have? Fill in the table.

I	Sister	Brother
7	6	13

? 13	7+6=13
7 6	Answer: 13 cookies.

3. Draw the missing shape in the box and write in their names.

octagon heptagon hexagon pentagon quadrilateral

4. Find 3 triangles in the shape and write the letters, for example, CBD, in the box. The order of the letters does not matter.

```
    A
      C
  B       D
```
1. A B C
2. B C D
3. A B D

---

1. Number Sense Strategy. Complete each number sentence and fill in the missing numbers. The first one is done for you.

8 + 5   7 + 8   9 + 3
 2 3   2 5   1 2

tens	ones
1	
	8
+	5
1	3

tens	ones
1	
	7
+	8
1	5

tens	ones
1	
	9
+	3
1	2

8 + 5 = 8 + 2 + 3 =   7 + 8 = 8 + 2 + 5 =   9 + 3 = 9 + 1 + 2 =
10 + 3 = 13   10 + 5 = 15   10 + 2 = 12

2. Explain how we subtract using the pictures below.

18 - 9 = 18 - 8 - 1 = 9
         8 1

18

I would build a fortress, a castle, and an ARMY with these sticks!

?9 1 8

3. How many do you need to make 1 whole? Write in the missing fraction.

$\frac{1}{2} + \frac{1}{2}$ $\frac{3}{4} + \frac{1}{4}$ $\frac{5}{6} + \frac{1}{6}$ $\frac{2}{3} + \frac{1}{3}$ $\frac{9}{10} + \frac{1}{10}$ $\frac{4}{5} + \frac{1}{5}$ $\frac{7}{8} + \frac{1}{8}$

---

1. Sketch a picture of yourself in the box. Take a measuring tape and measure yourself. Write down the measurements: Answers will vary.

- Standing height ....
- Sitting height ....
- Head length ....
- Neck circumference ....
- Right shoulder ....
- Left shoulder ....
- Waist breadth ....
- Shoulder to elbow ....
- Forearm hand length ....
- Foot length ....
- Knee circumference ....

I am _____.

Chest breadth (under the armpits) with exhalation: ....
Chest breadth (under the armpits) with inhalation: ....

Did you notice any difference? Why is the chest breadth different with inhalation and exhalation?

When you breathe in, the air comes into your lungs, and the chest breadth increases.

1. I've cut the 10-inch log into 2 equal pieces. How long is 1 part?
How many cuts did I make? 1.

Log	Equal pieces	Length of 1 part
10	2	?5

Answer: 5in., 1 cut.

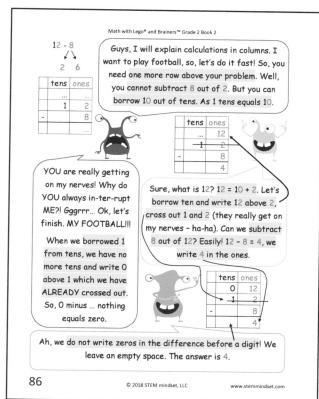

Ah, we do not write zeros in the difference before a digit! We leave an empty space. The answer is 4.

---

1. **Number sense Strategy.** **Complete** each number sentence and **write in** the missing numbers. The first one is done for you.

11 - 2   11 - 4   11 - 7
 1 1   1 3   1 6

tens	ones
0	11
~~1~~	~~1~~
-	2
	9

tens	ones
0	11
	1
-	4
	7

tens	ones
0	11
	1
-	7
	4

11 - 2 = 11 - 1 - 1 =   11 - 4 = 11 - 1 - 3 =   11 - 7 = 11 - 1 - 6 =
10 - 1 = 9   10 - 3 = 7   10 - 6 = 4

2. **Find** the value. **Remember:** 1 hour = 60 minutes.

```
  1 h 15 m       2 h 50 m       3 h 30 m
+ 1 h 15 m     + 4 h 20 m     + 1 h 35 m
  2 h 30 m       7 h 10 m       5 h 05 m
```

3. <u>Draw</u> the minute and hour hands red for midnight.
How much time is left until 25 past 9 am?
    9 hours 25 minutes
<u>Draw</u> the minute and hour hands black.

---

1. <u>Find</u> X by mental math. **Remember:** X = Whole = Part1 + Part2.

X - 7 = 6   X - 5 = 9   X - 9 = 2
X = 13   X = 14   X = 11

X - 4 = 8   X - 8 = 9   X - 6 = 9
X = 12   X = 17   X = 15

X - 9 = 9   X - 8 = 8   X - 7 = 8
X = 18   X = 16   X = 15

2. I turned 8 years old in 2017. When was I born? <u>Draw</u> the diagram.
2017: 8yo
Born: ? 2009

Turned	Year	Was born
8	2017	? 2009

2017 — 8 = 2009
Answer: 2009

3. There are 15 cards in my album. There are 10 cards less in my brother's album. How many cards do we have altogether?

My	Brother's	Brother's	Altogether
15	10 less	?5	?20

1: 15   15 - 10 = 5   2: ?20   15 + 5 = 20
   10   Answer: 5   15   5   Answer: 20
?5   cards.   cards.

<u>Circle</u> the correct number sentence: 15 + (15+10) or (15+10)+15
(15 + (15-10) or (15-10)+15.)

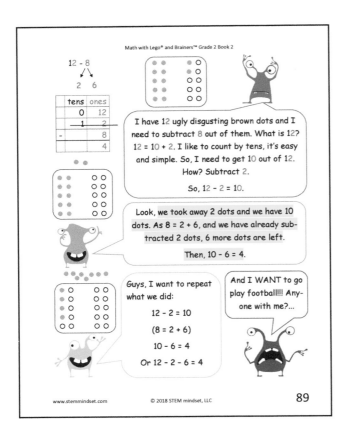

1. Number Sense Strategy. Complete each number sentence and write in the missing numbers.

11 - 9   12 - 3   12 - 8
 1   8   2   1   2   6

tens	ones
0	11
1	1
-	8
	2

tens	ones
0	12
1	2
-	3
	9

tens	ones
0	12
1	2
-	8
	4

11 - 9 = 11 - 1 - 8 =   12 - 3 = 12 - 2 - 1 =   12 - 8 = 12 - 2 - 6
10 - 8 = 2   10 - 1 = 9   = 10 - 6 = 4

2. Draw the minute and hour hands red for 2pm. How much time is left until 15 past 11 pm?
   9 hours 15 minutes
Draw the minute and hour hands black.

3. Measure 1 cup, 1 pint, 1 quarter, and 1 gallon of water. Find the connections between these measurements:

1 pint = 2 cups   1 quart = 4 cups   1 gallon = 16 cups
1 quart = 2 pints   1 gallon = 8 pints   1 gallon = 4 quarts

1 gallon = 4 quarts = 8 pints = 16 cups

---

1. Choose 2 weeks for weather observation. Check the sky at 11am and 6pm and fill in the table. Answers will vary.

The sky may be: Clear, Cloudy: 1/4, Cloudy: 1/2, Cloudy: 3/4, Cloudy
Precipitation may be: Rain, Thunder with lightning, Snow
The wind may be: North, East, South, West

Day	Sky		Wind		t°		Precipitation	
	11am	6pm	11am	6pm	11am	6pm	11am	6pm

---

1. Answer the questions. Answers will vary.
How many clear days did you check? _____
How many cloudy days did you check? _____
What kind of cloudy days did you observe the most? _____
How many windy days were there? _____
How many rainy days were there? _____
How many days had thunderstorms? _____
Circle the correct word: more/less and answer the questions.
How many more/less clear days than rainy days did you have?
_____.
How many more/less windy days than rainy days did you have?
_____.

Draw the columns for different kinds of days.

Project Weather: Days

14
12
10
8
6
4
2
0
   Rainy days   Snowy days   Clear days   Cloudy days   Windy days

Page 93

1. I've done a lot of art projects on Christmas vacation: 11 coloring pages, 34 oil paintings, 21 pencil drawings, and 9 Christmas 3D structures. How many art projects have I done in all?

   ?75

Coloring	Paintings	Drawings	3D structures	In all
11	34	21	9	?75

   | 11 | 34 | 21 | 9 |

   11+34+21+9 = (11+34) + (21+9) = 45+30=75

   Answer: 75 art projects.

2. Explain how we subtract using the pictures below.

   $13 - 9 = 13 - 3 - 6 = 4$

   3   6

   13   ?4   6   3

3. Find 3 rectangles in the shape and write the letters, for example, FKDB, in the box.

   ```
   A     C
   F     K
   B     D
   ```

   1. A C K F
   2. F B D K
   3. A B D C

   Write the letters clockwise or counter-clockwise, but not diagonally☺.

Page 94

1. I go to bed at 9 pm and wake up at 7 am. Dad goes to bed later and wakes up at 6 am. When I asked him when he went to bed, he said he was sleeping 2-and-a-half hours less than me. When does he usually go to bed? Draw the diagrams for the problem.

	I go to bed	I wake up	I sleep	Dad wakes up	Dad sleeps	Dad goes to bed
	9 pm	7 am	?10h00m	6 am	7h30m	?10:30 pm

   3 + 7 = 10 hours (I sleep)
   10:00 − 2:30 = 7:30 or 7 hours 30 minutes (Dad sleeps)
   7:30 − 6 = 1:30; 12:00 − 1:30 = 10:30 pm (Dad goes to bed)

   Answer: He goes to bed at 10:30 pm.

2. I've cut the 10-inch chain into 2 equal pieces. How long is 1 part?

   How many cuts did I make? 2.

Chain (in)	Equal pieces	Length of 1 part (in)
10	2	?5

   | 10 |
   | ?5 | ?5 |

   5+5=10

   Answer: 5 in., 2 cuts.

Page 95

1. Open any book or workbook and answer the questions.

   What page (p.) is before p. 9? p. 8
   What page (p.) is after p. 14? p. 13
   Can you find 2 consecutive numbers of pages if their sum is 13?
                                                p. 6 and p. 7
   Can you find 2 consecutive numbers of pages if their sum is 16?
   Why yes or no? No, the sum must be an odd number.

2. Number Sense Strategy. Fill in the missing numbers, write the number sentences, and find the value.

   8 + 8   8 + 8 = 16
   / \   8 + 2 + 6 = 16
   2 6

   9 + 3   9 + 3 = 12
   / \   9 + 1 + 2 = 12
   1 2

   8 + 3   8 + 3 = 11
   / \   8 + 2 + 1 = 11
   2 1

   6 + 6   6 + 6 = 12
   / \   6 + 4 + 2 = 12
   4 2

3. What bills and coins may make up $6.85? Answers will vary.

   Each quantity of bills and coins should be different.
   … 5-dollar bills, … 1-dollar bills, … quarters, … dimes, … nickels, … pennies.
   … 5-dollar bills, … 1-dollar bills, … quarters, … dimes, … nickels, … pennies.

Page 96

1. Number Sense Strategy. Complete each number sentence and write in the missing numbers.

   18 + 3   17 + 9   19 + 3
   / \   / \   / \
   2 1   3 6   1 2

tens	ones
1	
1	8
+	
	3
2	1

tens	ones
1	
1	7
+	
	9
2	6

tens	ones
1	
1	9
+	
	3
2	2

   18 + 3 = 18 + 2 + 1 =   17 + 9 = 17 + 3 + 6   19 + 3 = 19 + 1 + 2
   20 + 1 = 21   = 20 + 6 = 26   = 20 + 2 = 22

2. There were 17 sunny days in September, cloudy days were 9 less. How many cloudy days were in September? The remaining days were rainy. How many rainy days did we have?

Sept.	☀	☁	🌧
?30	17	9 less	?8	?5

   1: 17
      ?8   9

   17−9=8
   Answer: 8 cloudy days.

   2: 30
      17 8 ?5

   30−17−8=5
   Answer: 5 rainy days.

3. Who can hide in the 3-feet high grass? Find as many answers as you can imagine: Answers will vary.

Page 97

1. I have written the number 11. My sister has written the number which is 17 more. What number has she written? Draw the diagram.

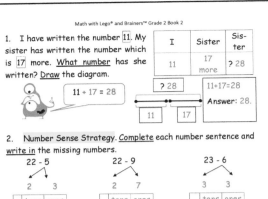

2. Number Sense Strategy. Complete each number sentence and write in the missing numbers.

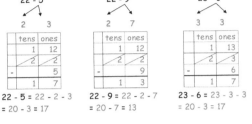

22 - 5 = 22 - 2 - 3
= 20 - 3 = 17

22 - 9 = 22 - 2 - 7
= 20 - 7 = 13

23 - 6 = 23 - 3 - 3
= 20 - 3 = 17

3. Compare the longest and the shortest sides of the shapes. Write the difference in the box.

2 - 1 = 1 in.   3 - 1 = 2 in.

Page 98

1. Check the scale balances. If you find a mistake, write the correct number in the parenthesis.

9 (11) — 11   16 - 7 (11) — 6 + 4   5 (8) — 12 - 4

3 (…) — 14 - 9   13 - 8 (8) — 7   17 (23) — 3 + 19

10 + 9 (…) — 18   18 - 7 (10) — 5 + 5   15 (11) — 6 + 6

2. Explain how we subtract using the pictures below.

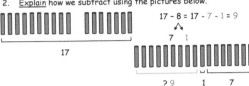

17 - 8 = 17 - 7 - 1 = 9

17   ? 9   1   7

3. Find 5 rectangles in the shape and write the letters, for example, LNKC, in the box. Answers may vary.

1. A L N F   2. F B D K
3. L N K C   4. A B D C
5. A F K C

Page 99

1. Compare the longest and the shortest sides of the shapes. Write the difference in the box. Find 1 mistake and cross it out.

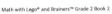

2 - 1.5 = 0.5 in.   3.5 - 3 = 0.5 in.

3 - 0.5 = 2.5 in.   … - … = … in.

2. We had some rose bushes in the backyard. Mom bought 3 more and now we have 11 rose bushes. How many bushes did we have before?

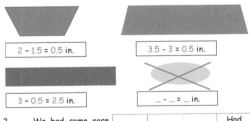

Had	Bought	Have	Had had
some	3	11	? 8

? = X

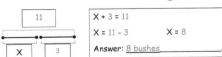

X + 3 = 11
X = 11 - 3   X = 8
Answer: 8 bushes

3. How can a giant of 10-feet high come into your room? Imagine as many ways as you can: Answers will vary

Page 100

1. Both of my Grandpas breed bees and harvest honey. Last year Mom's Pa made 32 lb. of honey, and my Father's Pa made 9 lb. more. How many lb. of honey did they make in all?

Mom's Pa	Dad's Pa	Dad's Pa	In all
32	9 more	? 41	? 73

1: ? …   32+9=41   2: ?73   41+32=73
   Answer: 41 lb.   Answer: 73 lb.

Circle the correct one number sentence for the problem:

32 + 32 + 9   (32 + 9) + 32

2. Number Sense Strategy. Fill in the missing numbers, write the number sentences, and find the value. Use Round-Up-Strategy.

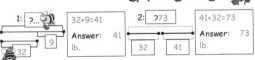

19 + 7   19 + 7 = 26   17 + 4   17 + 4 = 21
  /\   19 + 10 - 3 = 26   /\   17 + 10 - 6 = 21
10 3   10 6

15 + 7   15 + 7 = 22
  /\   15 + 10 - 3 = 22
10 3

3. Draw 2 lines with a ruler so that the square will be divided into 4 equal triangles.

Page 101

1. <u>Find</u> the value.

$\frac{2}{3} + \frac{3}{3} = \frac{5}{3}$ $\frac{1}{3} + \frac{2}{3} = \frac{3}{3}$ $\frac{2}{3} + \frac{2}{3} = \frac{4}{3}$

$\frac{1}{3} + \frac{1}{3} = \frac{2}{3}$ $\frac{1}{3} + \frac{4}{3} = \frac{5}{3}$ $\frac{3}{3} + \frac{3}{3} = \frac{6}{3}$

$1 + \frac{3}{3} = 1\frac{3}{3}$ $3 + \frac{1}{3} = 3\frac{3}{3}$ $2 + \frac{2}{3} = 2\frac{2}{3}$

1/3					

2. You need the scales and an apple. The color or size do not matter☺.

<u>Take</u> an apple and <u>weigh</u> it, and <u>write down</u> the weight. Then, <u>cut</u> it into 4 pieces and <u>leave</u> the pieces under the sun (or in the oven, but you need the help of an adult and cannot do it on your own) to dry it out. When it's dried out, <u>take</u> the dried pieces together and <u>weigh</u> them. <u>Write down</u> the weight.

Weight before: _____   Weight after: _____

<u>What</u> did you notice? Weight after is less_____.

<u>How</u> can you explain it? Since the water dried out and evaporated, the weight became less_____.

3. <u>Find</u> the value. <u>Remember</u>: 1 hour = 60 minutes.

```
  4 h 20 m       5 h 40 m       2 h 15 m
+ 2 h 15 m     + 2 h 30 m     + 7 h 45 m
  6 h 35 m       8 h 10 m      10 h 00 m
```

Page 102

1. <u>Number Sense Strategy</u>. <u>Complete</u> each number sentence and <u>write in</u> the missing numbers.

16 + 5 (4, 1)   17 + 4 (3, 1)   19 + 5 (1, 2)

tens	ones
1	
1	6
+	5
2	1

tens	ones
1	
1	7
+	4
2	1

tens	ones
1	
1	9
+	5
2	4

16 + 5 = 16 + 4 + 1 = 20 + 1 = 21

17 + 4 = 17 + 3 + 1 = 20 + 1 = 21

19 + 5 = 19 + 1 + 4 = 20 + 4 = 24

2. <u>Draw</u> the minute and hour hands red for 7:05pm. I usually go to sleep 1 hour and 45 minutes later.

<u>Draw</u> the minute and hour hands black for when I go to bed. 7:05 + 1:45 = 8:50pm____.

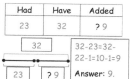

3. I had 23 bricks. <u>How many bricks</u> have I added if I have 32 bricks now?

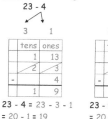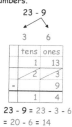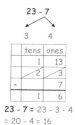

Had	Have	Added
23	32	? 9

32	
23	? 9

32-23=32-22-1=10-1=9

Answer: 9.

Page 103

2. <u>Number Sense Strategy</u>. <u>Complete</u> each number sentence and <u>write in</u> the missing numbers.

23 - 4 (3, 1)   23 - 9 (3, 6)   23 - 7 (3, 4)

tens	ones
1	13
2	
	4
1	9

tens	ones
1	13
2	
	9
1	4

tens	ones
1	13
2	
	7
1	6

23 - 4 = 23 - 3 - 1 = 20 - 1 = 19

23 - 9 = 23 - 3 - 6 = 20 - 6 = 14

23 - 7 = 23 - 3 - 4 = 20 - 4 = 16

2. <u>Take</u> a ruler and a good mood to answer the questions.

I need rectangles of 1 cm long and 3 cm wide. If I have a sheet of paper which is 9 cm long and 6 cm wide, <u>how many rectangles</u> can I cut out?

18 rectangles.

<u>Draw</u> the dotted lines on the rectangle to show your cutting lines.

3. <u>Continue</u> the pattern:

1, 6, 11, 16, 21, 26, 31   72, 69, 66, 63, 60, 57, 54

Page 104

"<u>Draw</u> the minute and hour hands red for 9:30am. The clock is 5 minutes behind or slow." Hm… Behind? Slow? The clock is hiding behind WHAT?! Don't you think it's scary? Definitely, creepy… Oh, I am so scared of math.

Calm down! Nothing scary! Sometimes the clock goes faster, sometimes it goes slower. It's a kind of broken clock, that's all! If you cannot fix it, just remember: Imagine that you are walking and I am walking behind you or I am slower. What do I need to do to catch you? Go faster and add more steps.

Do you mean I need to add the minutes "behind" to the time?

As 9:30 + 0:05 = 9.35?

Perfect! When the clock is slow or behind, you just need to add the minutes to the time on the clock and you get the right time!

1. Draw the minute and hour hands red for 3:15 pm.
The clock is 10 minutes behind or slow.
What time is it now?   3 : 25 pm___.
Draw the minute and hour hands black.

2. Write the number words.
458 Four hundred fifty-eight _____
746 Seven hundred forty-six _____
_____
583 Five hundred eighty-three _____
_____
837 Eight hundred thirty-seven _____
_____

3. I had some bricks. My sister added 8 bricks, and now I have 26 bricks. How many bricks did I have? Draw the scheme for the problem.

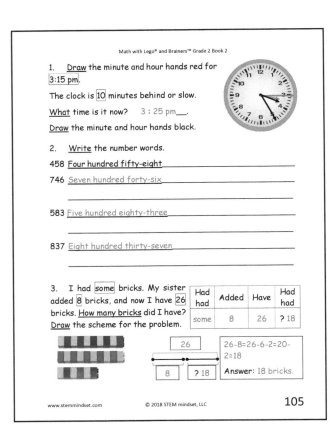

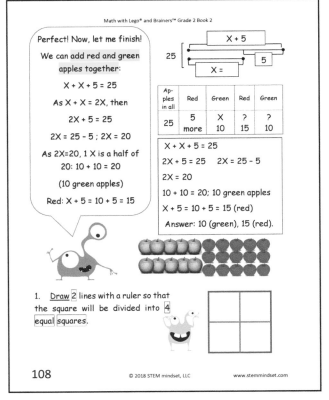

1. Draw 2 lines with a ruler so that the square will be divided into 4 equal squares.

Page 109

1. Last month we started a swim team. There were 42 kids, and the boys were 12 more than the girls. How many boys were on the team? How many girls were on the team?

Kids in a team	Boys	Girls	Boys	Girls
42	12 more	X	? 27	? 15

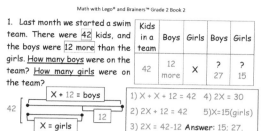

1) $X + X + 12 = 42$ 4) $2X = 30$
2) $2X + 12 = 42$ 5) $X = 15$ (girls)
3) $2X = 42 - 12$ Answer: 15; 27.

2. I put 3 sticks to show the race track. The interval between sticks was 20 steps. How many steps did I take?

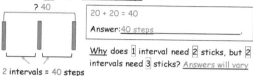

$20 + 20 = 40$
Answer: 40 steps.

2 intervals = 40 steps

Why does 1 interval need 2 sticks, but 2 intervals need 3 sticks? Answers will vary

3. I decided to measure the length of my room with my steps. I counted 15 steps. My big brother counted 11 steps. My Dad counted 9 steps. How can it be? We used the steps but got different numbers: The size of the steps was different for each person

What would you use to measure the room? Measuring tape.

Page 110

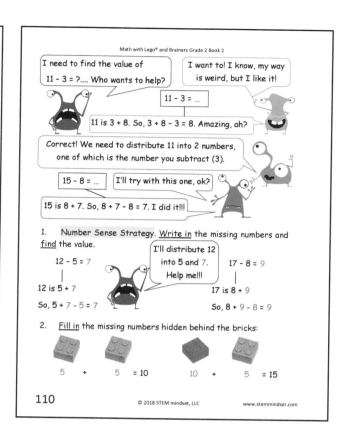

1. Number Sense Strategy. Write in the missing numbers and find the value.

$12 - 5 = 7$
12 is $5 + 7$
So, $5 + 7 - 5 = 7$

$17 - 8 = 9$
17 is $8 + 9$
So, $8 + 9 - 8 = 9$

2. Fill in the missing numbers hidden behind the bricks:

$5 + 5 = 10$ $10 + 5 = 15$

Page 111

1. Number Sense Strategy. I added two addends to 7 and got 26. Fill in the missing numbers in the number sentences to make them true. Answers will vary.

$7 + ... + ... = 26$ $7 + ... + ... = 26$
$7 + ... + ... = 26$ $7 + ... + ... = 26$

Cross out or circle 7 bricks and play with the rest!

2. Number Sense Strategy. Fill in the missing numbers, write the number sentences, and find the value.

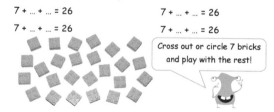

$19 + 6$ → 1, 5 $19 + 6 = 25$ $19 + 1 + 5 = 25$
$19 + 7$ → 1, 6 $19 + 7 = 26$ $19 + 1 + 6 = 26$
$19 + 5$ → 1, 4 $19 + 5 = 24$ $19 + 1 + 4 = 24$
$16 + 5$ → 4, 1 $16 + 5 = 24$ $16 + 1 + 4 = 24$

3. Draw the minute and hour hands red for 5:30 am. The clock is 15 minutes behind/slow.

What time is it now? 5 : 45 am.

Draw the minute and hour hands black.

Page 112

1. We bought peaches in a box. We ate 9 peaches in the evening. 16 peaches were left. How many peaches did we buy?

Peaches	Ate	Left	Buy
box	9	16	?25

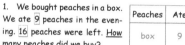

$9 + 16 = 25$
Answer: 25 peaches.

2. I need a gallon of water in each pot. How many more cups do I need to add to each pot to equal one gallon? Fill in each box.

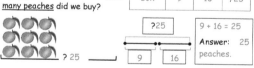

1 gallon = 16 cups   5 cups + 11 cups   8 cups + 8 cups
9 cups + 7 cups   4 cups + 12 cups

3. I had 45 bricks. How many bricks have I added if I have 75 bricks in all?

Had	Have	Added
45	75	? 30

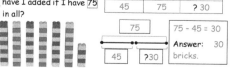

$75 - 45 = 30$
Answer: 30 bricks.

Math with Lego® and Brainers Grade 2 Book 2

1. We bought tomatoes in a box. We put 7 tomatoes in the salad. 14 tomatoes were left. How many tomatoes did we buy?

Toma-toes	Salad	Left	Buy
some	7	14	?21

?21

7+14=21

Answer: 21 tomatoes.

? 21 | 7 | 14

2. Number Sense Strategy. Fill in the missing numbers and find the value. Use Round-Up-Strategy.

21 − 8 = 13
|
21 is 10 − 2
So, 21 − 10 + 2 = 13

24 − 7 = ...
|
24 is 10 − 3
So, 24 − 10 + 3 = 17

3. Draw the minute and hour hands red for 7:45 am. The class starts in 25 minutes.
What time will the bell ring?   8:10 am___
Draw the minute and hour hands black.

4. The shape was turned 2 times to the left. Color the rectangles in their new position.

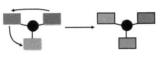

113

Math with Lego® and Brainers Grade 2 Book 2

1. There were 25 episodes in the movie. I have watched 7 episodes. How many episodes were left unwatched?

Were	Watched	Left
25	7	?18

were
watched   ? left

25
7 | ?18

25−7−18
Answer: 18 episodes.

2. There were several episodes in the movie. I have watched 17 episodes and 8 more were left. How many episodes were there?

Were	Watched	Left	Were
some	17	8	?25

? were
watched   left

?25
17 | 8

17+8=25
Answer: 25 episodes.

3. Number Sense Strategy. Explain how we subtract using the pictures below.

16 − 9 = 16 − 6 − 3 = 7
            6 3

16

? 7   3   6

114

Math with Lego® and Brainers Grade 2 Book 2

1. Number Sense Strategy. Fill in the missing numbers and write the number sentences. Find the value.

28 + 7   28 + 7 = 35
 ∧   28 + 2 + 5 = 35
2 5

29 + 4   29 + 4 = 33
 ∧   29 + 1 + 3 = 33
1 3

38 + 4   38 + 4 = 42
 ∧   38 + 2 + 2 = 42
2 2

27 + 6   27 + 6 = 33
 ∧   27 + 3 + 3 = 33
3 3

2. Fill in the missing numbers.

equals $\frac{5}{8}$ as   equals $\frac{2}{3}$

equals $\frac{3}{4}$ as   equals $\frac{3}{6}$

3. Open any book or workbook and answer the questions.
What page (p.) is before p. 18?   p. 17
What page (p.) is after p. 25?   p. 26
Can you find 2 consecutive numbers of pages if their sum is 19?
                                        p. 9 and p. 10
Can you find 2 consecutive numbers of pages if their sum is 14?
Why yes/no? No, the sum must be an odd number_____.

115

Math with Lego® and Brainers Grade 2 Book 2

1. Number Sense Strategy. I added three addends to 5 and got 30. Fill in the missing numbers in the number sentences to make them true. Answers will vary.

5 + ... + ... + ... = 30   5 + ... + ... + ... = 30
5 + ... + ... + ... = 30   5 + ... + ... + ... = 30
5 + ... + ... + ... = 30   5 + ... + ... + ... = 30

I circle 5 bricks and arrange the leftover bricks in any order.

2. There are 17 miles between my home and the swimming pool. We have already driven several miles and 9 miles are left. How many miles have we already driven?

Miles in all	Have already driven	Are left	Have already driven
17	several	9	? 8

17
X | 9

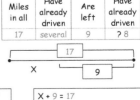

?

17
X | 9

X + 9 = 17
X = 17 − 9   X = 8
Answer: 8 miles.

116

Page 117

1. **Number Sense Strategy.** Explain how we subtract using the pictures below.

15

$15 - 7 = 15 - 5 - 2 = 8$

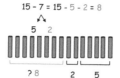
? 8   2   5

2. **Number Sense Strategy.** Fill in the missing numbers and write the number sentences. Find the value. Use Round-Up-Strategy.

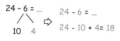

24 - 6 = ...   24 - 6 = ...   31 - 5 = ...   31 - 5 = 26
  10 4   24 - 10 + 4 = 18   10 5   31 - 10 + 5 = 26

3. Write any 2 consecutive numbers.   ..., ...
Write any 3 consecutive numbers.   ..., ..., ...
Write any 4 consecutive numbers.   ..., ..., ..., ...
Write any 5 consecutive numbers.   ..., ..., ..., ..., ...

4. Find the value.

```
  1
  6 h 50 m        8 h 10 m         1 h 32 m
+ 2 h 10 m      +   h 39 m       + 4 h 34 m
  9 h 00 m        8 h 49 m         6 h 06 m
```

Page 118

1. I've cut the 12-inch log into 2 equal pieces. How long is 1 part?

How many cuts did I make? 1.

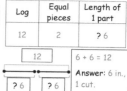

Log	Equal pieces	Length of 1 part
12	2	? 6

12 | 6 + 6 = 12
? 6 | ? 6 | Answer: 6 in., 1 cut.

2. **Number Sense Strategy.** Complete each number sentence and write in the missing numbers. Use Round-Up-Strategy.

28 + 8   9 + 49   4 + 39
 10 2   10 1   10 6

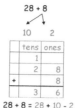
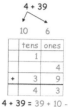

tens	ones
1	
	2	8
+		8
	3	6

tens	ones
1	
		9
+	4	9
	5	8

tens	ones
1	
		4
+	3	9
	4	3

$28 + 8 = 28 + 10 - 2$
$= 38 - 2 = 36$

$9 + 49 = 49 + 10 - 1$
$= 59 - 1 = 58$

$4 + 39 = 39 + 10 - 6 = 49 - 6 = 43$

3. Find the value.

```
    1 $ 3 6 ¢      6 $ 2 2 ¢       ...
                                 7 $ 4 5 ¢
  + 9 $ 1 2 ¢    + 2 $ 2 7 ¢    + 1 $ 2 5 ¢
   10 $ 4 8 ¢      8 $ 4 9 ¢      8 $ 7 0 ¢
```

Page 119

1. I had bricks. I shared 9 bricks with my friend and now I have 15. How many bricks did I have before sharing?

Had	Shared	Have	Had
bricks	9	15	? 24

X - 9 = 15
X = 9 + 15   X = 24
Answer: 24 bricks_____.

2. **Number Sense Strategy.** Explain how we subtract using the pictures below.

14

$14 - 8 = 14 - 4 - 4 = 6$
       4 4

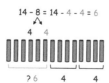
? 6   4   4

3. I want to buy a book for $9.50, and a set toys for $6. I found in my piggy bank: 2 5-dollar bills, 11 1-dollar bills, 12 quarters, 5 dimes, 4 nickels, and 100 pennies.

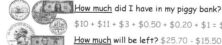

How much did I have in my piggy bank?

$10 + $11 + $3 + $0.50 + $0.20 + $1 = $25.70.

How much will be left? $25.70 - $15.50 = $10.20.

Page 120

1. **Number Sense Strategy.** Fill in the missing numbers and write the number sentences. Find the value. Use Round-Up-Strategy.

26 - 9 = 17   21 - 6 = 15
   | |
26 is 10 - 1   21 is 10 - 4

So, 26 - 10 + 1 = 17   So, 21 - 10 + 4 = 15

2. I had some games on the iPad. I added 9 games, and now I have 28 total. How many games did I have before?

Had	Added	Have	Had
some	9	28	? 19

28

X   9

X + 9 = 28
X = 28 - 9   X = 19
Answer: 19 games_____.

3. Answer the questions.

How many ends are on 1 log of a tree?   2.
How many ends are on 3 logs?   6.
How many ends are on 2 logs and a half?   6.

Page 121

1. Number Sense Strategy. Complete each sentence and fill in the missing numbers. Use Round-Up-Strategy.

24 - 5   34 - 7   44 - 9
10 5   10 3   10 1

tens	ones
	14
2	4
-	5
1	9

tens	ones
2	14
3	4
-	7
	7

tens	ones
3	14
4	9
-	9
3	5

24 - 5 = 24 - 10 + 5 = 14 + 5 = 19

34 - 7 = 34 - 10 + 3 = 24 + 3 = 27

44 - 9 = 44 - 10 + 1 = 34 + 1 = 35

2. Explain how we subtract using the pictures below.

12 - 7 = 12 - 2 - 5 = 5

3. Find the missing of 2 consecutive numbers. 53, 54.
Find the missing of 3 consecutive numbers. 26, 27, 28.
Find the missing of 4 consecutive numbers. 9, 10, 11, 12.
Find the missing of 5 consecutive numbers. 60, 61, 62, 63, 64.

Page 122

1. Number Sense Strategy. I added two addends to 9 and got 29. Fill in the missing numbers in the number sentences to make them true. Answers may vary.

9 + ... + ... = 29
9 + ... + ... = 29
9 + ... + ... = 29
9 + ... + ... = 29
9 + ... + ... = 29

2. Find the sum or difference.

400 + 200 = 600   270 + 510 = 780   705 + 100 = 805
637 + 300 = 937   300 + 520 = 820   150 + 440 = 590
490 - 220 = 270   680 - 160 = 520   275 - 130 = 145
957 - 320 = 637   759 - 340 = 419   407 - 100 = 307

3. Answer the question.
Draw the minute and hour hands red for 11:40 am. The science class lasts 45 minutes.
When will it be over? 12:25 pm____.
Draw the minute and hour hands black.

4. Continue the pattern: 4, 8, 12, 16, 20, 24, 28, 32, 36.

Page 123

1. I've counted 85 roses on 2 bushes. The red roses were 25 more than the pink roses. How many red roses were there? How many pink roses were there?

Roses in all	Red	Pink	Red	Pink
85	25 more	X	? 55	? 30

X + 25 = red
85
X = pink
25

X + X + 25 = 85
2X = 85 - 25
2X = 60   X = 30
X + 25 = 55
Answer: 55 red roses; 30 pink roses.

X + 25 = 55   X = 30
85

2. Find 8 triangles in the shape and write the letters, for example, BKD, in the box. The order of the letters does not matter.

1. A B K   2. A K C
3. A B C   4. B K D
5. B C F   6. A B D
7. A B F   8. A D F

3. Divide the rectangle into equal 4 parts where each part has the same number of squares and 1 circle.

Page 124

1. I am as old as my sister and my sister is older than my brother. My cousin is older than my sister. Draw the diagram to show the age of kids in our family from the oldest to the youngest. (For example, I am older than my brother show as I > Brother).

Cousin > Sister = I > Brother

2. Answer the questions.
Sketch a circle in the box and divide it into 5 equal parts. Compare the fractions, using <, >, or =:

$\frac{1}{5} < \frac{4}{5}$? $\frac{3}{5} > \frac{2}{5}$?

3. Number Sense Strategy. Complete each sentence and fill in the missing numbers. Use Round-Up-Strategy.

27 + 6   6 + 49   5 + 77
10 4   10 4   10 5

tens	ones
	1
2	7
+	6
3	3

tens	ones
	1
	6
+ 4	9
5	5

tens	ones
	1
	5
+ 7	7
8	2

27 + 6 = 27 + 10 - 4 = 37 - 4 = 33

6 + 49 = 49 + 10 - 4 = 59 - 4 = 55

5 + 77 = 77 + 10 - 5 = 87 - 5 = 82

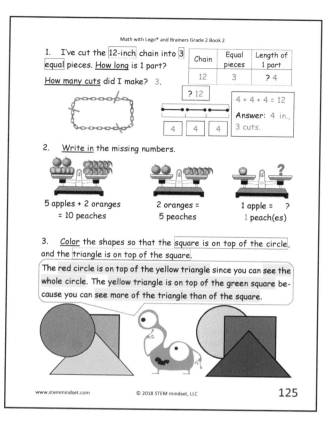
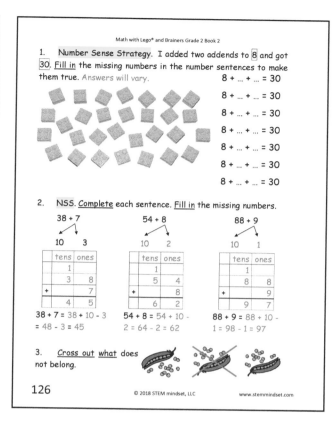
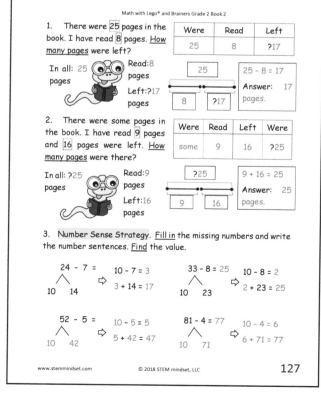

Page 129

1. <u>Number Sense Strategy</u>. <u>Fill in</u> the missing numbers and <u>write</u> the number sentences. <u>Find</u> the value.

47 + 6	47 + 6 = 53	78 + 6	78 + 6 = 84
∧	47 + 3 + 3 = 53	∧	78 + 2 + 4 = 84
3 3		2 4	

29 + 8	29 + 8 = 37	59 + 5	59 + 5 = 64
∧	29 + 1 + 7 = 37	∧	59 + 1 + 4 = 64
1 7		1 4	

2. You need a scale, 3 tablespoons of salt, a cup of water and a large shallow saucer or a plate, and a good mood ☺.

Take 3 tablespoons of salt, weigh it, and write down the weight.

Then, dissolve it into a half cup of water, pour it onto a large shallow saucer or a plate, leave it under the sun to evaporate (or ask for an adult to help you boil it in a pot until it's evaporated). When all the water is evaporated, pick up the leftover salt and weigh it. Write down the weight. Answers will vary.

Weight before: _____ Weight after: _____

What did you notice? _____

How can you explain it? _____

_____

Page 130

1. <u>Explain</u> how we subtract using the pictures below.

11 − 6 = 11 − 1 − 5 = 5

2. I had 7 red bricks. Then, I added 14 blue and 9 green bricks. How much more did I add to what I had?

Had	Ad-dends	Added	Added > had
7	14, 9	?23	?16

1: ?23   14 + 9 = 23   2: 23   23 − 7 = 16
   Answer: 23 bricks.   Answer: 16 more.
14 9   7   ?16

3. <u>Find</u> the missing of 2 odd consecutive numbers. 43, 45.
<u>Find</u> the missing of 3 odd consecutive numbers. 13, 15, 17.
<u>Find</u> the missing of 4 odd consecutive numbers. 17, 19, 21, 23.
<u>Find</u> the missing of 5 odd consecutive numbers. 31, 33, 35, 37, 39.

4. The shape was turned 2 times to the right. <u>Color</u> the rectangles in their new position.

Page 131

1. <u>Write</u> the number sentence for the problem. <u>Find</u> the value.

83621: <u>Find</u> the sum of the thousands and tens 3 + 2 = 5___, the difference of the hundreds and ones 6 − 1 = 5_____, the sum of the ten thousands, hundreds and tens 8 + 6 + 2 = 16, the difference of the ten thousands, thousands and ones 8 − 3 − 1 = 4_____.

2. I've counted 14 sailors on 2 boats. The white boat had 2 more sailors than the blue boat. <u>How many sailors</u> were on the white boat? <u>How many sailors</u> were on the blue boat?

Sailors in all	White	Blue	White	Blue
14	2 more	X	? 8	? 6

14 { X + 2 = white
     X = blue }   2

X + X + 2 = 14
2X = 14 − 2   2X = 12
X = 6   6 + 2 = 8
Answer: 6 blue, 8 white.

3. <u>Follow</u> the directions and <u>color</u> the circles.

<u>Start</u> with the brown circle; <u>go</u> 3 circles to the right and 3 circles down. <u>Color</u> the circle green.

Then, <u>go</u> 2 circles to the left, 2 circles up, and 3 circles to the right. <u>Color</u> the circle yellow.

Page 132

1. <u>Number Sense Strategy</u>. <u>Complete</u> each sentence and <u>fill in</u> the missing numbers.

34 − 8	45 − 7	25 − 9
4 4	5 2	5 4

tens	ones
2	14
3	4
-	
	8
2	6

tens	ones
3	15
4	5
-	
	7
3	8

tens	ones
1	15
2	5
-	
	9
1	6

34 − 8 = 34 − 4 − 4   45 − 7 = 45 − 5 − 2   25 − 9 = 25 − 5 − 4
= 30 − 4 = 26   = 40 − 2 = 38   = 20 − 4 = 16

2. <u>Number Sense Strategy</u>. I added two addends to 6 and got 31. <u>Fill in</u> the missing numbers. Answers will vary.

6 + ... + ... = 31   6 + ... + ... = 31
6 + ... + ... = 31   6 + ... + ... = 31

6 + ... + ... = 31

Do you know, TANGRAM which means seven boards of skill, was supposed to be invented in China and brought to Europe in the 19th century? Cut out the colored shapes and be ready to play!

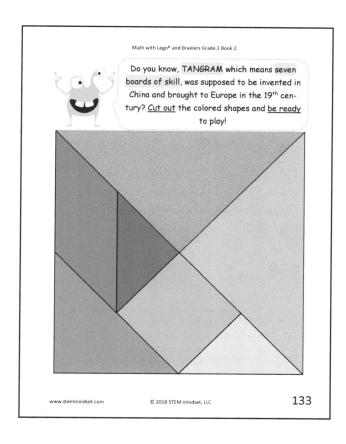

133

134

You can build different shapes with the pieces of this amazing puzzle. Let's try with the colored shapes.

135

1. I had 9 green bricks, then, I added 7 yellow and 5 red bricks. How much more did I add to what I had? Draw the diagrams.

Had	Ad-dends	Added	Added > had
9	7, 5	? 12	? 3

1: ?12   7+5=12   2: 12   12-9=3
   7 5   Answer: 12 bricks.   9 ?3   Answer: 3 more.

2. Number Sense Strategy. Fill in the missing numbers and find the value. The first one is done for you.

12 − 5 = 7   17 − 8 = 9
   ↓
12 is 5 + 7   17 is 8 + 9

So, 5 + 7 − 5 = 7   So, 8 + 9 − 8 = 9

3. My 2 sisters (Anna and Mary) and I (Lisa) were playing "Scary Woods". We chose 3 creatures: a dragon, a fairy, and a princess. Mary and I were not princesses. Mary is also a romantic, she even sleeps with a magic wand. Find out what creatures we chose. Write Y for Yes and N for No.

	Dr.	F.	Pr.
A.	N	N	Y
M.	N	Y	N
L.	Y	N	N

4. Find X. Remember: Part1 = Whole − Part2.

X + 200 = 700   X + 400 = 900   X + 100 = 700
X = 500   X = 500   X = 600

136

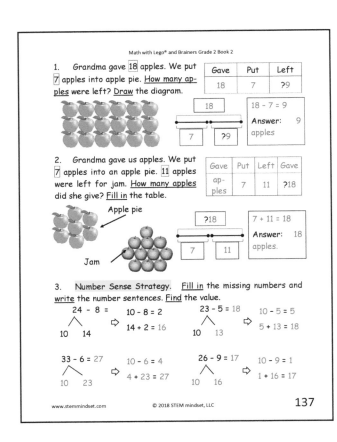

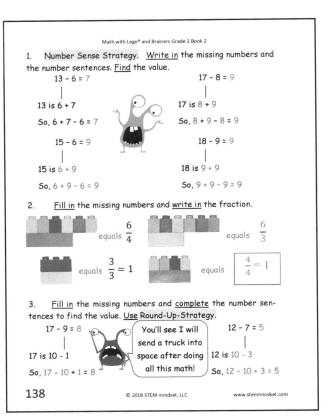

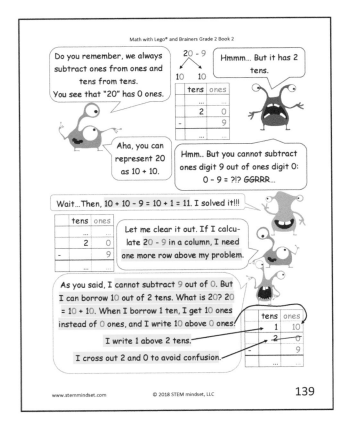

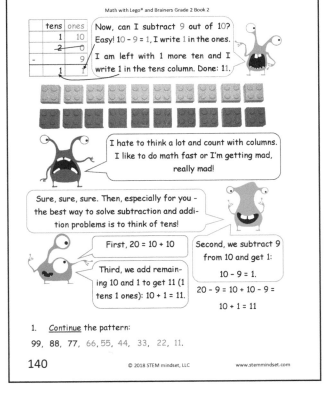

1. Continue the pattern:

99, 88, 77, 66, 55, 44, 33, 22, 11.

Math with Lego® and Brainers Grade 2 Book 2

1. I need 30 bricks of orange and blue colors each. If I have 20 orange bricks and 19 blue bricks, how many more bricks of each color do I need?

Need orange and blue each	Have orange	Have blue.	Need orange	Need blue
30; 30	20	19	?10	?11

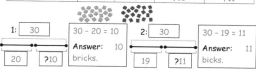

1: 30   30 − 20 = 10   2: 30   30 − 19 = 11
20 ?10   Answer: 10   19 ?11   Answer: 11
             bricks.   bicks.

2. Find the missing addend. Remember: Whole = Part1 + Part2.

X − 3 = 47   X − 2 = 78   X − 8 = 52
X = 50   X = 80   X = 60
? − 4 = 26   ? − 6 = 44   ? − 9 = 81
? = 30   ? = 50   ? = 90
🐵 − 5 = 65   🦁 − 7 = 23   🐱 − 1 = 39
🐵 = 70   🦁 = 30   🐱 = 40

3. Compare the fractions, using >, <, or =.

$\frac{5}{6}$ > $\frac{2}{3}$ $\frac{4}{6}$ = $\frac{2}{3}$ $\frac{3}{3}$ = $\frac{6}{6}$

1/3		
1/6			

141

Math with Lego® and Brainers Grade 2 Book 2

1. Number Sense Strategy. Complete each number sentence, fill in the missing numbers. Use your favorite strategy. Answers vary.

46 − 7   25 − 8   35 − 6
 ... .

tens	ones
3	16
4	6
−	7
3	9

tens	ones
1	15
2	5
	8
1	7

tens	ones
2	15
3	5
	6
2	9

46 − 7 = _____   25 − 8 = _____   35 − 6 = _____

2. Find 4 triangles in the shape and write the letters, for example, FCD, in the box. Answers will vary.

A — — — — — C
| F |
B — — — — — D

1. A C F
2. C F D
3. A B D
4. A C D

3. Find X. Remember: Whole = Part1 + Part2.

X − 200 = 300   X − 400 = 600   X − 100 = 700
X = 500   X = 1000   X = 800

4. Cross out what does not belong.

142

Math with Lego® and Brainers Grade 2 Book 2

1. Find the value.

$1 - \frac{2}{3} = \frac{1}{3}$ $1 - \frac{1}{3} = \frac{2}{3}$ $1 - \frac{3}{3} = \frac{0}{3} = 0$

$\frac{2}{3} - \frac{1}{3} = \frac{1}{3}$ $\frac{4}{3} - \frac{1}{3} = \frac{3}{3} = 1$ $\frac{3}{3} - \frac{2}{3} = \frac{1}{3}$

$\frac{3}{3} + \frac{2}{3} - \frac{1}{3} = \frac{2}{3}$ $\frac{4}{3} - \frac{2}{3} - \frac{1}{3} = \frac{1}{3}$ $\frac{2}{3} + \frac{3}{3} - \frac{4}{3} = \frac{1}{3}$

2. Take a ruler, measure the rectangle, and answer the questions.

Divide a rectangle into 6 equal parts.

How many is 1 part of the rectangle? $\frac{1}{6}$.

What kinds of shapes did you get?

Squares or rectangles.

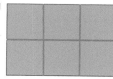

3. My Grandma has chickens and rabbits. I've counted 11 heads and 32 legs.

How many chickens does she have? 6 chickens = 12 legs

How many rabbits does she have?

5 rabbits = 20 legs;

6 + 5 = 11 heads; 12 + 20 = 32 legs

143

Math with Lego® and Brainers Grade 2 Book 2

"Draw the minute and hour hands red for 9:30am. The clock is 5 minutes ahead or fast." Hm… Ahead? The clock is WHAT?! Mad, I am soooo mad!!! Why can't the clock work right? What's the problem with the clock? Let's stop doing math and take the clock to the repair shop, right?

No, no, no! Nothing to be mad about. Imagine: you are walking and I am walking ahead of you or I am faster.

I hate it when you are walking faster, I need to walk more and make more steps to catch you or ask you to slow down or freeze. All options drive me mad.

Correct! When the clock is fast or ahead, you just need to subtract the minutes out of the time on the clock and you get the right time: 9:30 − 0:05 = 8:25

1. Cross out the picture which does not belong.

144

1. Circle the bricks which you need to fill up the rectangle and cross out the bricks which are not needed.

Answers will vary.

2. NSS. Complete the number sentences and fill in the missing numbers.

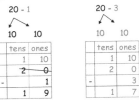

20 - 1 = 10 + 10 - 1 = 10 + 9 = 19

20 - 3 = 10 + 10 - 3 = 10 + 7 = 17

20 - 5 = 10 + 10 - 5 = 10 + 5 = 15

3. Compare fractions using the bars below, insert >, <, or =.

$\frac{1}{2}$ < $\frac{4}{6}$ $\frac{1}{6}$ < $\frac{1}{3}$ $\frac{2}{3}$ = $\frac{4}{6}$

$\frac{1}{6}$					
$\frac{1}{3}$					
$\frac{1}{2}$					

---

1. Find the missing of 2 even consecutive numbers. 26, 28.
Find the missing of 3 even consecutive numbers. 16, 18, 20.
Find the missing of 4 even consecutive numbers. 28, 30, 32, 34.
Find the missing of 5 even consecutive numbers. 42, 44, 46, 48, 50.

2. I had $76 from my birthday. I've spent $57. How much more have I spent than is left over?

	Had	Spent	Left over	Spent > Left
	76	57	?19	?38

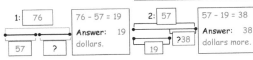

3. Fill in the missing numbers so that each side of the triangle must add up to the number in the center.

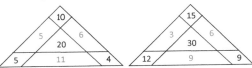

4. Find X. Remember: Part1 = Whole – Part2.

X + 125 = 678   X + 375 = 595   X + 425 = 575
X = 553   X = 220   X = 150

Now you can try to make the black shapes with the TANGRAM puzzle pieces.

---

1. It was 47°F in the morning and 55°F in the afternoon. How much warmer was it in the afternoon?

Morning	Afternoon	How much warmer
47	55	?8

55 - 47 = 8
Answer: 8 degrees.

2. It was 47°F in the morning, it got 5° more in the afternoon. How warm was it in the afternoon?

Morning	After-noon	Afternoon
47	5 more	?52

47 + 5 = 52
Answer: 52 degrees.

3. School starts at 7:45am. Draw the minute and hour hands red. Mom dropped me off at 7:15am. Draw the minute and hour hands black. How long did I wait for the first class?

30 minutes.

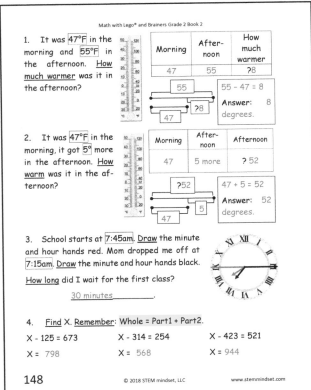

4. Find X. Remember: Whole = Part1 + Part2.

X - 125 = 673   X - 314 = 254   X - 423 = 521
X = 798   X = 568   X = 944

Page 149

1. There are 30in. between a guitar and a chair in my little sister's castle. How many inches will be between them if I move the chair 15 inches to the right and the guitar 17 in. to the left? Draw the arrows to show my moves.

?62

30 + 15 + 17 = 62
Answer: 62 inches.

| 30 | 15 | 17 |

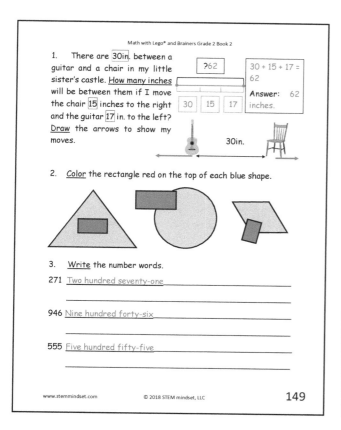

2. Color the rectangle red on the top of each blue shape.

3. Write the number words.

271 Two hundred seventy-one

946 Nine hundred forty-six

555 Five hundred fifty-five

Page 150

1. I've drawn a 10cm long and 8cm high rectangle. You may look at what Briner says below☺.

Find the Perimeter:

P = 10 + 10 + 8 + 8 = 36

Find 3 different measurements of rectangle's length and height if the Perimeter is 36 sq.cm., the length is more than the height:

Length = 12 cm   Length = 11 cm   Length = 15 cm
Height = 6 cm   Height = 7 cm   Height = 3 cm

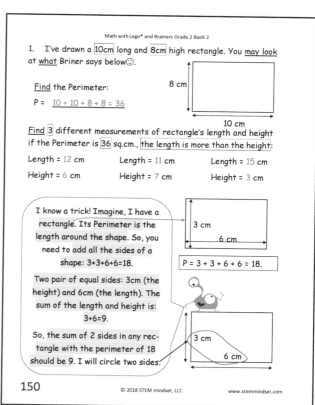

I know a trick! Imagine, I have a rectangle. Its Perimeter is the length around the shape. So, you need to add all the sides of a shape: 3+3+6+6=18.

Two pair of equal sides: 3cm (the height) and 6cm (the length). The sum of the length and height is: 3+6=9.

So, the sum of 2 sides in any rectangle with the perimeter of 18 should be 9. I will circle two sides:

P = 3 + 3 + 6 + 6 = 18.

Page 151

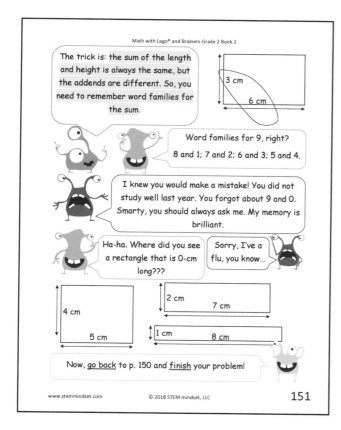

The trick is: the sum of the length and height is always the same, but the addends are different. So, you need to remember word families for the sum.

Word families for 9, right? 8 and 1; 7 and 2; 6 and 3; 5 and 4.

I knew you would make a mistake! You did not study well last year. You forgot about 9 and 0. Smarty, you should always ask me. My memory is brilliant.

Ha-ha. Where did you see a rectangle that is 0-cm long???

Sorry, I've a flu, you know...

Now, go back to p. 150 and finish your problem!

Page 152

1. Compare fractions using the bars below, fill in >, <, or =.

$\frac{1}{3} = \frac{2}{6}$ $\frac{4}{6} < \frac{3}{3}$ $\frac{1}{3} < \frac{1}{2}$

$\frac{5}{6} > \frac{3}{6}$ $\frac{6}{6} = 1$ $\frac{3}{6} = \frac{1}{2}$

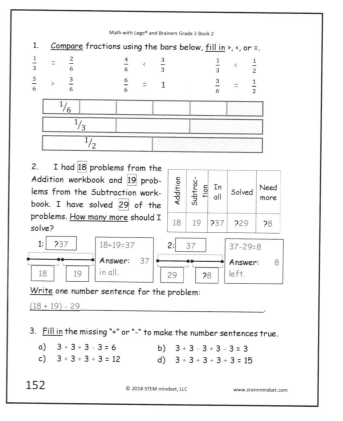

2. I had 18 problems from the Addition workbook and 19 problems from the Subtraction workbook. I have solved 29 of the problems. How many more should I solve?

Addition	Subtraction	In all	Solved	Need more
18	19	?37	?29	?8

1: ?37   18+19=37   2: 37   37-29=8
| 18 | 19 | Answer: 37 in all. | 29 | ?8 | Answer: 8 left.

Write one number sentence for the problem:
(18 + 19) - 29

3. Fill in the missing "+" or "-" to make the number sentences true.
a) 3 + 3 + 3 - 3 = 6   b) 3 + 3 - 3 + 3 - 3 = 3
c) 3 + 3 + 3 + 3 = 12   d) 3 + 3 + 3 + 3 + 3 = 15

1. Now you can try to make the black shapes.

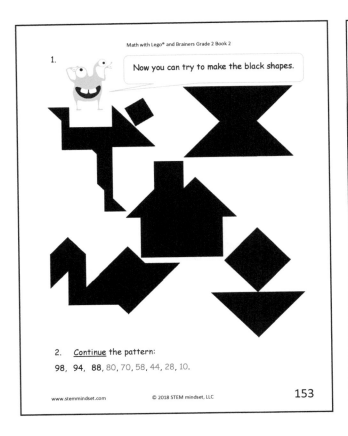

2. Continue the pattern:
98, 94, 88, 80, 70, 58, 44, 28, 10.

---

1. Follow the directions and color the circles.

Start with the blue circle: go 4 circles to the right, 2 circles down, and 3 circles to the left. Color the circle red.

Then, go 1 circle up, 2 circles to the right, 2 circles down, 3 circles to the left. Color the circle green.

2. I want to buy a chocolate ice cream for $3.50, and a toy for $7.99. I found in my piggy bank: 1 5-dollar bills, 4 1-dollar bills, 20 quarters, 8 dimes, 20 nickels, and 19 pennies.

How much did I have in my piggy bank?
$5.00+$4.00+$5.00+$0.80+$1.00+$0.19=$15.99
How much will be left? $3.50 + $7.99 = $11.49
$15.99 - $11.49 = $4.50.

3. I have 95 bricks: 22 are red and 31 are green. How many yellow bricks do I have? Draw the scheme.

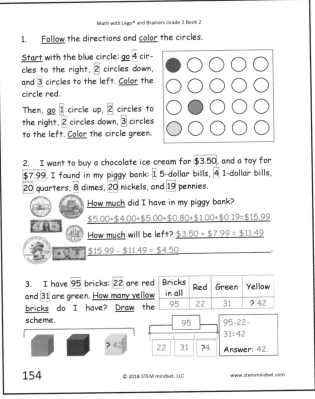

Bricks in all	Red	Green	Yellow
95	22	31	?42

95-22-31=42
Answer: 42.

---

1. Fill in the missing numbers and find the value.

12 - 4 = 8   13 - 7 = 6
| |
12 is 4 + 8   13 is 7 + 6
So, 4 + 8 - 4 = 8   So, 7 + 6 - 7 = 6

2. Check the scale balances. If you find a mistake, write the correct number in the parenthesis.

18(22) 39-17   25(20) 47-26   21(24) 19+5

44 (22) 60-38   18 (26) 56-31   32 (37) 27+9

45(44) 37+6   67 (50) 48+2   22(12) 30-17

3. I subtracted 5 and 7 bricks out of 33 bricks. How many more or less bricks are left over than were subtracted?

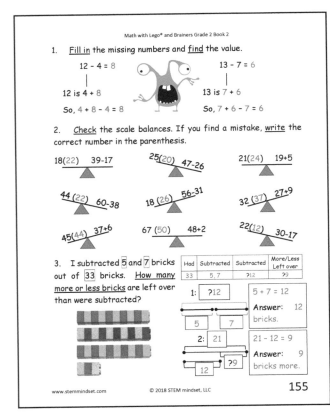

Had	Subtracted	Subtracted	More/Less Left over
33	5, 7	?12	?9

1: ?12   5 + 7 = 12
            Answer: 12 bricks.

2: 21   21 - 12 = 9
            Answer: 9 bricks more.

---

1. Guess the width and the length of your backyard (or school's playground, or your room). Guess the length as well. Sketch the shape in the box and write down your measuring guesses.

Measure the width and the length with a measuring tape, sketch the shape in the box, and write down the actual measurements.

Guess: Measurements:

Answers will vary.

What is the difference in the widths? ... - ... = ... _____.
What is the difference in the lengths? ... - ... = ... _____.
Use a timer to find out how fast you can walk from one end of the backyard (or a playground, or a room) diagonally to the opposite end?   ... : ....
Use a timer to find out how fast you can run from one end of the backyard (or a playground, or a room) diagonally to the opposite end?   ... : ....

Page 157

1. <u>Use</u> the clock to solve the problem.

I must be at school at ⟨7:30 am⟩ ☺. I have a ⟨30-minute⟩ lunch break at ⟨10.30 am⟩ and a ⟨30-minute⟩ recess at ⟨12 am⟩ ☺. I leave school at ⟨2:30 pm⟩. <u>How much time</u> do I study at school?

	Come to school	Lunch break	Recess	Leave school	Time to study in school
	7:30	30min 10:30	30min 12:00	2:30	? 6:00

10:30-7:30=3:00   12:00-11:00=1:00
2:30-12:30=2:00   3:00 + 1:00 + 2:00 = 6:00

Answer: <u>6 hours</u>

2. My little sister likes to play tea party with her dolls. She puts them <u>around the table</u>. Her Ariel (A) doll is always sitting the fifth from the Rapunzel (R) doll, <u>clockwise or counter-clockwise</u>. <u>How many dolls</u> are sitting at the table? <u>Draw</u> the dolls. <u>10 dolls</u>

Page 158

1. I have built the tower with ⟨17⟩ bricks. My sister's tower has ⟨9 less⟩ bricks. <u>How many bricks</u> does she have?

I	Sister	Sister
17	9 less	? 8

 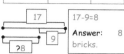

17-9=8
Answer: 8 bricks.

2. <u>Number Sense Strategy.</u> <u>Fill in</u> the missing numbers and <u>write</u> the number sentences. <u>Find</u> the value.

49 + 9 = 58   37 + 7 = 44
 ∧   ∧
1 8   49 + 1 + 8 = 58   3 4   37 + 3 + 4 = 44

3. I added ⟨5⟩, ⟨8⟩, and ⟨3⟩ to ⟨4⟩. <u>How much more</u> is the sum now than the number ⟨4⟩?

Had	Addends	Sum	Sum > had
4	5, 8, 3	?20	?16

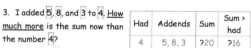

1: ?20   4+5+8+3=20   2: 20   20-4=16
4 5 8 3   Answer: 20   4   ?16   Answer: 16 more.

4. <u>Find</u> X. Remember: Part1 = Whole – Part2.

X + 300 = 760   X + 200 = 580   X + 400 = 850
X = 460   X = 380   X = 450

Page 159

1. <u>Number Sense Strategy.</u> <u>Complete</u> each number sentence and <u>fill in</u> the missing numbers to find the value. <u>Use</u> Round-Up-Strategy.

30 - 7   40 - 2   50 - 4
 ∧   ∧   ∧
10 3   10 8   10 6

tens	ones
2	10
3	0
2	3

tens	ones
3	10
4	0
3	8

tens	ones
4	10
5	0
4	6

30 - 7 = 30 - 10   40 - 2 = 40 - 10   50 - 4 = 50 - 10
+ 3 = 20 + 3 = 23   + 8 = 30 + 8 = 38   + 6 = 40 + 6 = 46

2. <u>Color</u> the triangle yellow so that it's underneath of a green shape.

3. I have $⟨55⟩. My brother has $⟨7⟩ ⟨more⟩. <u>How many dollars</u> does he have?

I have	Brother	Brother
55	7 more	? 62

Brother has ? $...
I've $...

 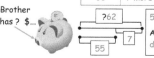

55+7=62
Answer: 62 dollars.

Page 160

1. <u>Number Sense Strategy.</u> <u>Fill in</u> the missing numbers and <u>write in</u> the number sentences. <u>Find</u> the value.

28 - 9 = 19   ⇒   10 - 9 = 1   25 - 7 = 18   ⇒   10 - 7 = 3
 ∧   1 + 18 = 19   ∧   3 + 15 = 18
10 18   10 15

37 - 9 = 28   ⇒   10 - 9 = 1   32 - 8 = 24   ⇒   10 - 8 = 2
 ∧   1 + 27 = 28   ∧   2 + 22 = 24
10 27   10 22

2. If my sister and I add our candies, we will have ⟨24⟩ candies. If we subtract them, the difference will be ⟨4⟩ candies. She has ⟨more⟩ candies than I have. No wonder, she's older☺. <u>How many candies</u> do I have?

Sum	Difference	Sister	I
24	4	more	? 10

10 + 14 = 24
14 - 10 = 4

Answer: I have 10 candies, sister has 14 candies.

3. <u>Find</u> the total number of quadrilaterals and triangles of any size in these shapes.

  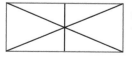

Quadrilaterals <u>2 + 1 = 3</u>
Triangles   <u>6 + 2 + 4 = 12</u>

1. Find out what number is hiding behind X, ?, or an animal's face. Remember: X = Part1 + Part2. Add ones. Then, add tens.

X - 13 = 47   X - 12 = 78   X - 38 = 52
X = 13 + 47 = 60   X = 78+12=90   X = 52+38=90
? - 34 = 26   ? - 36 = 44   ? - 19 = 81
? = 26+34=60   ? = 44+36=80   ? = 81+19=100
🐝 - 45 = 35   🦁 - 67 = 23   🐼 - 71 = 19
🐝 = 35+45=80   🦁 = 23+67=90   🐼 = 19+71=90

2. My brother is 14 years old, I am 8 years old. How old will my brother be when I am as old as he is now? (Brother is weird, ah?).

My brother now	I am now	Difference in our ages	My brother in 6 years
14	8	?6	?20

1: [14] — [8] [?6]   14 - 8 = 6   Answer: 6 years.

8y.o. → 14y.o → ?20y.o

2: [?20] — [14] [6]   14 + 6 = 20   Answer: 20 years old.

Write one number sentence for the problem:
(14 − 8) + 14

3. Find X.
X - 300 = 560   X - 200 = 580   X - 400 = 350
X = 860   X = 780   X = 750

---

1. Number Sense Strategy. I added three addends to 6 and got 28. Fill in the missing numbers in the number sentences to make them true. Answers will vary.

6 + ... + ... + ... = 28   6 + ... + ... + ... = 28
6 + ... + ... + ... = 28   6 + ... + ... + ... = 28
6 + ... + ... + ... = 28   6 + ... + ... + ... = 28

2. Answer the questions.

Mom cooked 4 steaks for dinner. There were 2 fathers and 2 sons at the table. Each of them ate 1 steak and there was 1 steak left on the plate. Is it possible? Who were they? Draw them sitting at the table.

Where did 1 son or Dad disappear?..

Who are they? They are: A Grandfather, a father, and a son.

Lightning Source UK Ltd.
Milton Keynes UK
UKHW05f1001220518
322949UK00006B/57/P